Abdallah SABER

Installation of a STEMI thrombolysis network in emergency departments

Abdallah SABER

Installation of a STEMI thrombolysis network in emergency departments

ScienciaScripts

Cover image: www.ingimage.com

This book is a translation from the original published under ISBN 978-620-6-70826-1.

Publisher:
Sciencia Scripts
is a trademark of
Dodo Books Indian Ocean Ltd. and OmniScriptum S.R.L publishing group

120 High Road, East Finchley, London, N2 9ED, United Kingdom
Str. Armeneasca 28/1, office 1, Chisinau MD-2012, Republic of Moldova, Europe
Managing Directors: Ieva Konstantinova, Victoria Ursu
info@omniscriptum.com

Printed at: see last page
ISBN: 978-620-8-37639-0

Contents

Signing sessions

In memory of my dearest father and my lovely mother-in-law, may God God welcome them into his vast paradise.

To my very dear mother and my very dear father-in-law, I wish them a long life.

I would like to thank all my family for their encouragement,

my very dear wife, whose support has been invaluable to me and who deserves all the credit for this work,

my dearest children for their love and patience,

my very dear brothers and swursfor their assistance.

To all my elders and masters whose work and effort have led to the transmission of my passion for anaesthesia and intensive care.

To the whole team at the Mostaganem hospital emergency department, and in particular the intensive care unit.

To all my colleagues, especially the intensive care anaesthetists and cardiologists.

CHAPTER I

INTRODUCTION - ISSUES

I. *INTRODUCTION :*

Coronary heart disease accounts for half of all cardiovascular deaths worldwide [1]. It is often associated with myocardial infarction (MI) [2]. MI is defined as cardiomyocyte necrosis in a clinical situation consistent with acute myocardial ischaemia [3]. MI is a serious manifestation of coronary artery disease, resulting from acute occlusion of a coronary artery by a thrombus, most often complicating the rupture or cracking of an atherosclerotic plaque [4,5].

According to the World Health Organisation (WHO), in middle-income countries including Algeria, mortality from ischaemic heart disease increased by more than 1 million between 2000 and 2019 [6]. It currently stands at 3.1 million [7].

Acute coronary syndrome (ACS) with ST-segment elevation (STEMI) is the most serious form of coronary artery disease, resulting from acute occlusion of one or more coronary arteries by thrombus from ruptured or cracked atherosclerotic plaque [8].

The prognosis of STEMI in the acute phase depends on the speed with which coronary flow is restored by angioplasty or thrombolysis. While angioplasty requires the availability of a catheterisation room and a permanently qualified team, thrombolysis is easier to carry out at the point of diagnosis and therefore remains the ideal treatment in our country for regions lacking suitable technical facilities. The installation of a STEMI thrombolysis network in the various regions of our country that do not have catheterisation rooms is more than necessary in order to overcome this real public health danger.

In Mostaganem, and due to a lack of permanent and stable cardiologists in the public sector, only seven patients with STEMI were identified and thrombolysed during the whole of 2018 (excluding December), and their morbidity and mortality was not specified due to a lack of traceability. Concerned, this situation of failure of care of STEMI in the region, is at the origin of our major concern since December 2018 or the need for the creation of a network of care of this pathology in the emergency department of Mostaganem proved necessary. As a result, resuscitators and cardiologists in Mostaganem have organised this network for better management of patients admitted for chest pain with ST-segment elevation on ECG less than 12 hours old in the emergency setting. These patients, recruited and listed on a register, will benefit from coronary reperfusion with drugs, prescribed and carried out by one of the two specialists and involving the general practitioner on duty, after all contraindications have been eliminated.

II. PROBLEM :

Coronary heart disease is a serious public health problem. According to the WHO, it is one of the leading causes of death worldwide, accounting for 29.6% of all deaths in 2010 and 26.7% in 2016 [1, 2].

In Algeria, the Tahina study clearly shows that the leading mortality rate in our country is represented by cardiovascular diseases, which accounted for more than 44.5% in 2002 [3], with an incidence of 41% in 2014 and 36% in 2016 [4, 5].

According to statements made by the President of the Algerian Society of Cardiology

(SAC) in 2013, cardiovascular pathologies, particularly MI, are responsible for more than 25,000 deaths a year, i.e. more than 2,000 deaths every month [6].
He also pointed out that, until 2013, our country had only 4,000 cases of coronary angioplasty per year, and only 25% of these were performed on an emergency basis (i.e. primary angioplasty and angioplasty for non-ST segment elevation ACS).
This lack of medical care is due to a shortage of staff and a limited number of catheterisation rooms (public-private) in just a few large towns. This regional disparity and the poor distribution of resources mean that almost 80% of the Algerian population does not have access to an angioplasty centre, as is the case in Mostaganem. Consequently, fibrinolytic treatment remains the main reperfusion therapy [7,8] for most patients with ST-segment elevation acute coronary syndrome (STEMI) in the different regions of our country.
In Mostaganem, the prognosis for STEMI remains serious, as patients often arrive late to be seen by cardiologists who are not on duty, given the limited number of cardiologists at the CHU.
Our UMC department handles all the wilaya's cardiovascular emergencies. On average, every month there are around ten cases of STEMI admitted to hospital as emergencies, not counting those that become complicated or die as a result of delayed or inadequate treatment. As a result, there is an urgent need to train general practitioners in Mostaganem's emergency services to diagnose and treat STEMI in order to improve the prognosis.
Our study will enable us to :

- To report our experience of thrombolysis of STEMI in the emergency department and discuss its benefits and side effects.
- To describe the epidemiological and clinical profile of patients admitted for ST+ ACS to the Mostaganem emergency department.
- Evaluate the delay in thrombolysis in relation to the onset of pain
- To record the results of thrombolysis and cardiovascular events during the hospital stay.
- To propose a thrombolysis strategy adapted to our region in order to improve the prognosis of STEMI.

CHAPTER II

Literature data

I. Acute coronary syndrome (ACS) :

A. Definitions :

1. Atherosclerosis :

According to the WHO, atherosclerosis is defined as a variable association of changes in the intima of large- and medium-calibre arteries, consisting of a local accumulation of lipids, complex carbohydrates, blood and blood products, fibrous tissue and calcareous deposits. All of this is accompanied by changes in the media [9,10].

Atherosclerosis is the main aetiology (95% of cases), but all CVDRFs are involved in the aetiopathogenesis of ACS [11].

2. Ischaemic heart disease :

They are secondary to complications of atherosclerosis in one or more coronary arteries and include lesions ranging from ischaemia to myocardial necrosis due to insufficient oxygen supply to the myocardium [12].

3. Acute coronary syndromes :

The classification of ACS is based on ST segment changes. This classification is established for operational purposes in order to distinguish immediately:

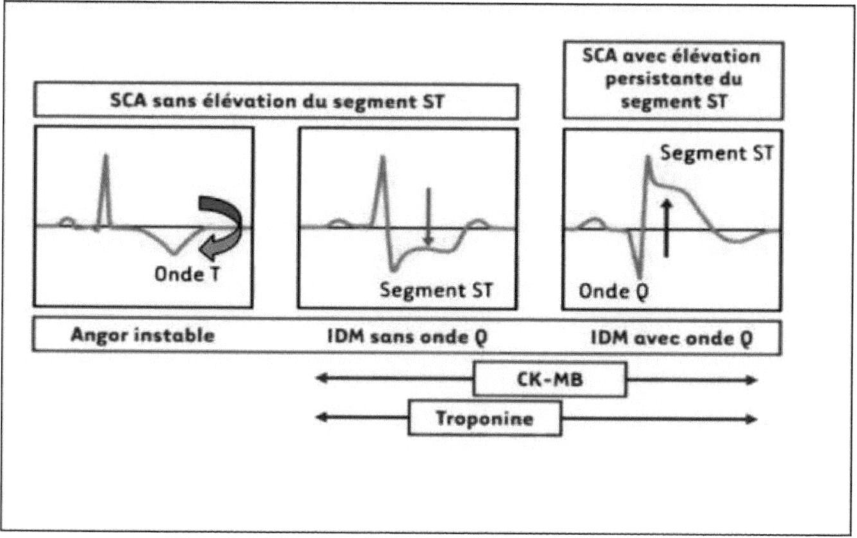

Figure 1: Nosology of ACS.

Persistent ST-segment elevation ACS (ST+ ACS) is associated with complete and sustained occlusion of an epicardial coronary artery, and requires **urgent revascularisation** using intravenous fibrinolysis or angioplasty.

Non-persistent ST-segment elevation ACS (ST-ACS), in which the coronary occlusion is partial or intermittent. Within SCAST-, a distinction is made between **non-Q-wave myocardial infarction** (or sub-endocardial **MI) and unstable angina pectoris**, depending on whether or not markers of myocardial necrosis are released. ST-ACS is a heterogeneous clinical entity, with an equally heterogeneous prognosis, and its management is based on risk stratification, enabling patients to be guided towards the most appropriate therapeutic strategy.

4. Acute coronary syndrome ST+ (ACS ST+) :

ACS ST+ corresponds to the acute myocardial infarction of the old definitions. It is defined on the ECG by persistent ST-segment elevation (STEMI).

Before 1999, MI was defined by the WHO as the combination of at least two of the following three criteria: a clinical history of chest pain lasting more than twenty minutes, changes on the ECG (electrocardiogram) and an increase then decrease in cardiac markers [13].

In 2000, the concept of ACS was introduced, which is better suited to emergency medicine and whose main symptom is acute chest pain, but patients are classified on the basis of the ECG. Two categories of patients can be distinguished:

- **Patients with acute chest pain and persistent (> 20 min) ST-segment elevation**: These are known as ST+ ACS (STEMI), which generally reflects an acute total coronary occlusion.
- **Patients with acute chest pain but without persistent ST-segment elevation: These are non persistent ST-segment elevation ACS (NSTEMI)** [14].

B. Anatomical reminder :

1-Arteries :

An artery (from the Greek artêria) is a blood vessel which, from a functional point of view, carries blood from the heart to other tissues in the body.

All the arteries carry oxygenated blood to the organs except the pulmonary arteries, which carry oxygen-poor blood to the lungs for re-oxygenation.

An artery is made up of several concentric layers:

- **The intima, in** direct contact with the blood, is made up of an endothelium (composed of squamous epithelial cells) and a subendothelial layer (loose connective tissue) separated by a basal lamina.
- **The media** is made up of smooth muscle cells surrounded by a extracellular matrix, itself made up of collagen and elastin fibres.
- The **adventitia** (periphery) is made up of collagen fibres, elastin and adipose cells (adipocytes).

2- Coronary arteries :

They are known as coronary arteries because they are arranged in a ring around the heart. There are two of them, **the right coronary artery (RCA)** and **the left coronary artery (LCA)** [15]. They arise from the initial portion of the aorta (see Figure 1). The left and right coronaries are **terminal arteries**. Their resting flow represents around 5% of cardiac output and, unlike most other arteries, they are perfused during ventricular diastole [16].

- **Left coronary artery**: the left coronary artery (arteria coronaria sinistra) arises from the sinus of the aorta (left coronary sinus). It is very short, 1 to 2 centimetres, between the pulmonary trunk and the left atrium. It divides into two branches: **an anterior interventricular artery** and **a circumflex artery**.
- **Right coronary artery:** the right coronary artery (arteria coronaria dextra) arises in the sinus of the aorta (sinus aortae) or right coronary sinus. Its course has three segments: a first short segment oblique forwards and to the right, a second segment which runs along the lower edge of the right atrium, then a third segment where it joins the posterior surface of the creur at the level of the sulci cross.
- It divides into two terminal branches: **the posterior interventricular artery** and the **left retroventricular artery**. Its collateral branches are right atrial ascending branches and right ventricular descending (marginal) branches [17].

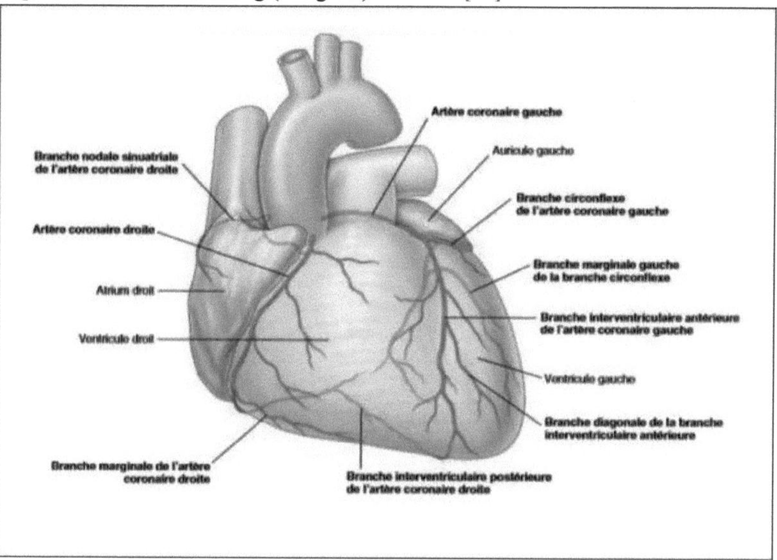

Figure 2: Anterior view of the coronary arterial system [18].

3-Vascular territories :

The left coronary artery vascularises the atrium, left atrium, left ventricle and the anterior 2/3 of the interventricular septum.

The right coronary artery vascularises the right atrium, the right auricle, the right ventricle, the inter-atrial septum and the posterior 1/3 of the inter-ventricular septum.

Thrombosis in an artery of the cardioverter system can lead to **rhythm disorders.** The vascularisation of the cardionic system includes:

- **The sinus na'iid**: is irrigated by the CD in 64% of cases and by the CG in 36% of cases;
- **The atrioventricular na'iid**: is supplied by the first posterior septal artery from the DC;
- **The atrioventricular bundle (AVB)**: also supplied by the first posterior septal artery

from the DC;

- **The right branch of the AVF**: supplied by the 2nd anterior septal artery of the GC;
- **The left branch of the AVF**: supplied by septal branches from the 2 coronaries.

C. Pathophysiological background :

1- Pathophysiology of atheromatous coronary artery disease [19,20]:

Atherosclerosis is a variable association of changes in the intima of large and medium-calibre arteries, consisting of a focal accumulation of **lipids**, **complex carbohydrates**, **blood** and **blood products**, **fibrous tissue** and **calcareous deposits**, accompanied by changes in the media.

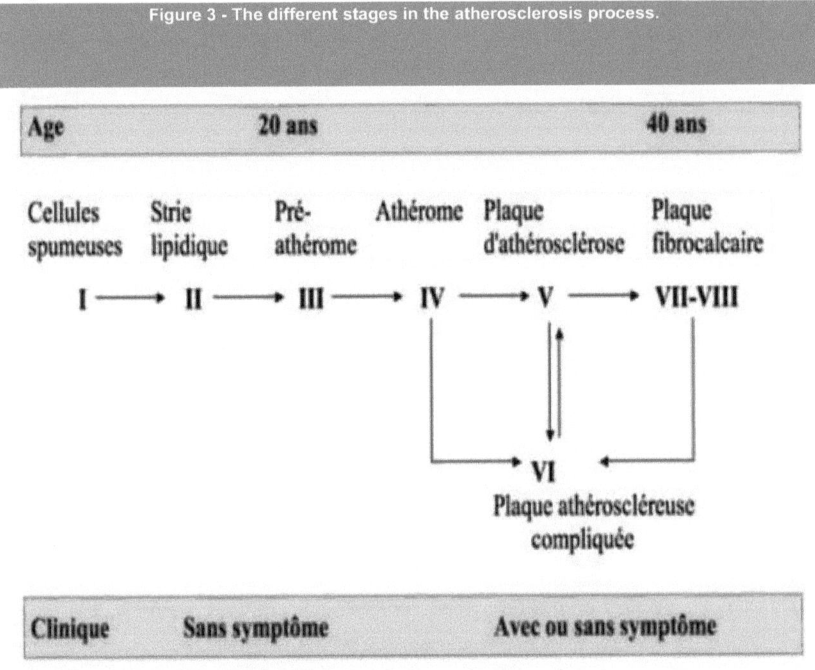

Figure 3: Genesis of atherosclerotic plaque [21].

rSuccessive mechanisms combine to form plaque:

- **Accumulation of low-density lipoproteins (LDL)** in the intima.

This is a passive phenomenon secondary to an imbalance between inputs and outputs.

- **Oxidation of LDL**: oxidation of LDL takes place in situ, in the intimal space, by various enzymatic and non-enzymatic mechanisms.
- **Endothelial dysfunction or activation**: Normal endothelium inhibits the adhesion of circulating monocytes. This activation leads to the expression of adhesion molecules on the endothelial surface and is largely dependent on the presence of oxidised LDL in the intima.
- **Recruitment of circulating monocytes and their transformation into macrophages and then into foam cells**: this phase involves circulating monocytes adhering to the surface of the endothelium, crossing it and transforming into macrophages and then into foam cells.

Macrophages present in the sub-endothelial space play a key role in many stages of atherosclerosis (production of inflammatory cytokines, etc.).

- **Chronic inflammatory reaction**: As soon as macrophages infiltrate the arterial wall, they cause a chronic inflammatory reaction which is crucial for plaque growth (self-amplification).

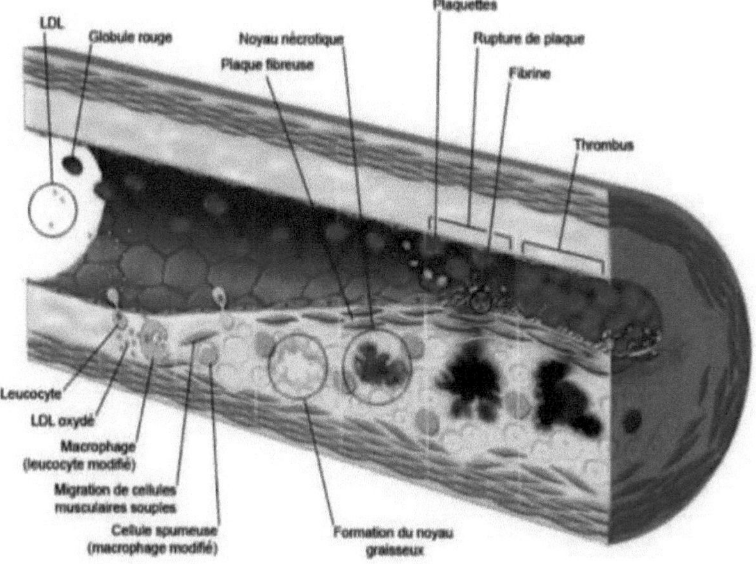

Figure 4: Formation of atherosclerotic plaque [22].

- **Formation of the atheromatous centre and the fibrous cap**: the lipids in the plaque are first essentially intracellular and then also extracellular; at this stage, they group together to form a mass known as the lipid creur or atheromatous centre. Adult atherosclerotic plaque is characterised by the formation of a fibromuscular cap which isolates the lipid centre from the arterial lumen. The fibrous cap is composed of smooth muscle cells and extracellular matrix proteins (collagen, elastin, proteoglycans). The smooth muscle cells come from the media. The integrity of the fibrous cap is a determining factor in the stability of atherosclerotic plaques.

2. Evolution of atherosclerotic plaque :

Atherosclerotic plaque rupture is a sudden complication, the cause of acute clinical accidents. At any moment, the inflammatory process may increase in the plaque and weaken the fibrous protein cap; this is the case when metalloproteases released by macrophages and smooth muscle cells digest the proteins of this protective fibrous shell and when locally secreted cytokines trigger apoptosis of the smooth muscle cells. The plaque is then liable to rupture, or to undergo erosion of its endothelial coating (unstable plaque), thereby exposing the tissue factor molecules it contains to the bloodstream. The blood will clot on contact with the tissue factor and the thrombus that forms may cause sufficient occlusion of the arterial lumen to trigger an acute ischaemic stroke. The thrombus may fragment and create emboli.

- **The progression of atheromatous plaque:** can lead to a reduction in the lumen of the vessel (lipid component and matrix). This progression may be gradual, but mainly occurs in flare-ups during acute plaque rupture accidents, incorporating thrombotic material. The plaque may slowly evolve into fibrous and calcified tissue.
- **Intraplaque haemorrhage**: leads to a sudden increase in the volume of the plaque and can rupture the fibrous cap.
- **Plate regression:** observed

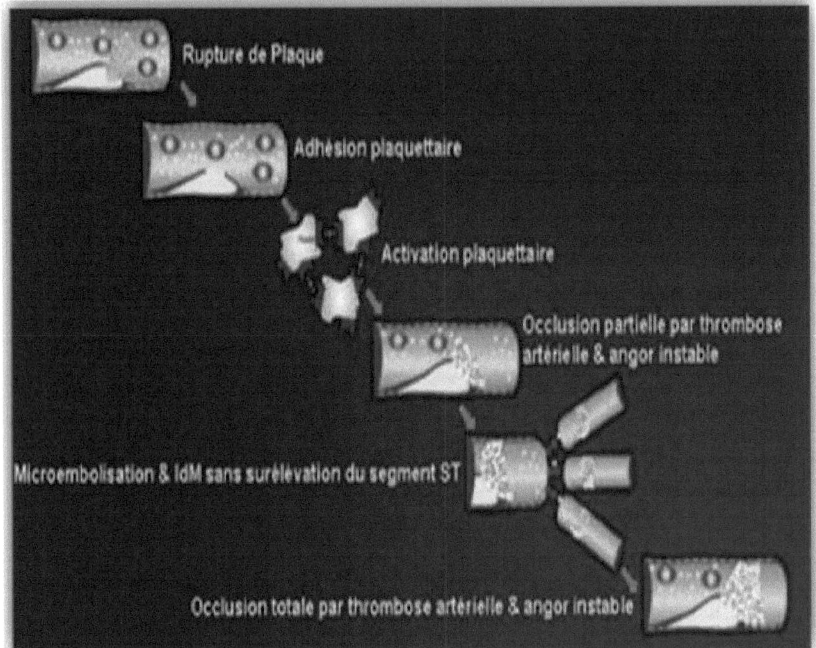

Figure 5 : Evolution of atherosclerotic plaque

3- Pathophysiology of acute coronary syndrome :

The mechanism of total coronary occlusion in STEMI is a thrombosis that develops on an atherosclerotic coronary plaque. Promoted by a combination of cardiovascular risk factors, coronary artery disease begins with the formation of an atherosclerotic plaque, the starting point of which is endothelial dysfunction. ACS has a common pathophysiological mechanism, represented mainly by plaque **rupture** or **erosion** of the coronary endothelium (observed above all in women). This plaque rupture or endothelial erosion activates platelet functions and the coagulation cascade, leading to the formation of **platelet** and/or **fibrino-cruciate thrombi** capable of abruptly reducing the vessel lumen or obliterating it completely.

In addition, thrombi formed at the site of plaque rupture or platelet erosion are likely to embolise into the distal bed and create single or multiple foci of cellular necrosis, explaining the enzymatic release [23].

4- Myocardial consequences of coronary thrombosis :

In the absence of treatment, and depending on the duration of the occlusion, the ischaemic threshold of the myocyte and whether or not there is bypass circulation, coronary thrombosis progresses towards myocardial ischaemia and then necrosis.

a. **Myocardial ischaemia:** This is the result of an imbalance between myocyte oxygen supply and consumption. This myocardial ischaemia lasts a maximum of 4 hours after the onset of coronary occlusion and leads to hypoxia or even cellular anoxia.

At the cellular level, abnormalities in myocyte sarcoplasm occur after around forty minutes in the myocardial subendocardium.

At a metabolic level, in the event of ischaemia, there is a deviation from the normal metabolic pathway towards anaerobic glycolysis. This pathway is a poor source of ATP and there is an accumulation of fatty acids which are harmful to the myocyte. In addition, the fall in coronary flow compromises the elimination of lactates and protons, resulting in metabolic acidosis. This secondary ATP deficit leads to dysfunction of transmembrane ion transport, with accumulation of intracellular calcium, and generates electrophysiological abnormalities leading to serious rhythm disorders [24,25].

Then, within a few minutes, contractile and haemodynamic abnormalities following coronary occlusion and hypokinesia will be seen opposite the ischaemic wall.

b. **Myocardial necrosis:** results from the perpetuation of acute thrombotic coronary occlusion. Transmural necrosis affects more than 50% of the myocardial thickness. According to experimental data, **necrosis is complete after** approximately **four hours**; in humans, **subendocardial myocardial necrosis begins thirty to forty minutes after the onset of occlusion**; it progresses from the subendocardium to the epicardium and from the centre to the periphery of the area concerned.

Histologically, the necrotic myocardium is colonised by eosinophilic polymorphs and monocytes. An inflammatory reaction appears within forty-eight hours and is constantly associated with this necrosis. Unlike ischaemia, this phase of necrosis is not reversible. Various parameters will influence the time taken for complete necrosis to develop: in particular, the existence or not of collaterals, the notion of preconditioning which corresponds to the repetition of phases of ischaemia-reperfusion preceding the infarction, and above all the therapeutic methods used at that precise moment [26,27].

c. **Systolic-diastolic dysfunction:** the deterioration in systolic function evidenced by the drop in ejection fraction is due to the amputation of the left ventricular contractile mass corresponding to the necrotic zone, which is akinetic or dyskinetic, and to a lesser extent to the reduction in contractility (hypokinesis) in the zones adjacent to the necrosis. On the other hand, the heterogeneity of contraction creates contractile asynchronism which aggravates ventricular dysfunction. The reduction in ventricular distensibility is linked to the prolongation of relaxation due to the ischaemia of the areas bordering the necrosis, resulting in an alteration in diastolic function.

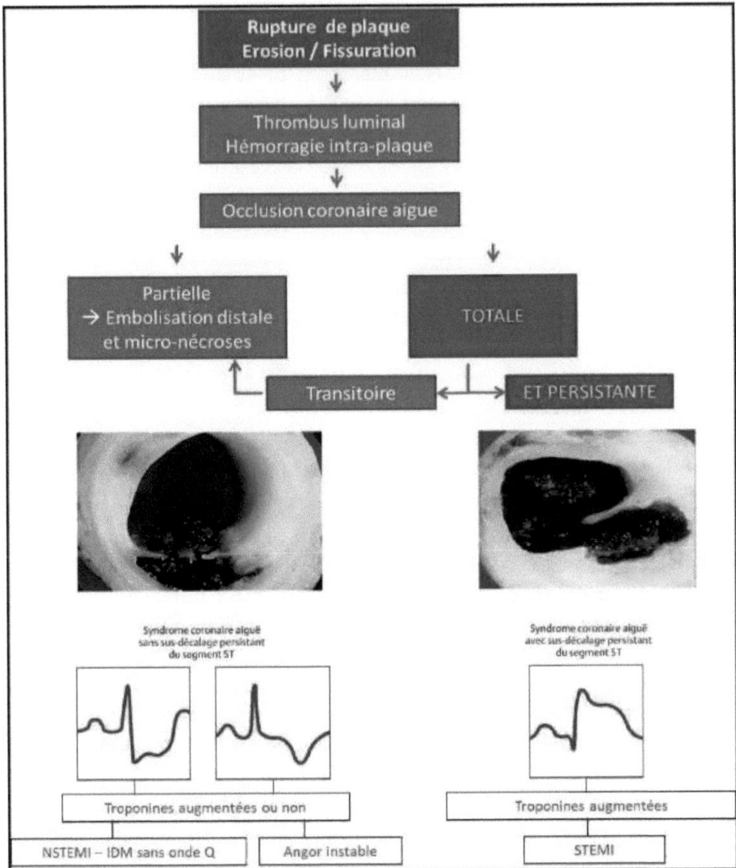

Figure 6: Pathophysiology of acute coronary syndromes.

d. Changes in ventricular geometry: These affect both the necrotic zone and the healthy myocardium. The thinning and dilation of the necrotic zone results in an expansion of this same zone, whereas the healthy myocardium undergoes hypertrophy-dilatation which is intended to be compensatory. Together, these two phenomena constitute **ventricular remodelling**. Factors predictive of this remodelling are the anterior topography of the infarct, an ejection fraction of less than 40%, an akinetic mass of more than 20% and the persistence of coronary occlusion. Corticosteroids and non-steroidal anti-inflammatory drugs have a detrimental effect on remodelling, and loading conditions also play a role, while angiotensin-converting enzyme inhibitors have the opposite effect **[28,29, 30].**

e. Biochemically: the death of myocytes leads to the release of myocardial enzymes into the bloodstream, which form the basis of the biological diagnosis of myocardial infarction.

f. Concept of reperfusion: the restoration of coronary flow in the few tens of minutes following coronary occlusion makes it possible to avoid myocardial necrosis.

Forty minutes, corresponding to the first sarcoplasmic lesions, would appear to be a crucial period, since within this time there is no necrosis on the electrocardiogram and, above all, no irreversible loss of contractile function. On the other hand, reperfusion attempted within the first four hours would marginally reduce necrosis, essentially in the border areas.

g. Sideration concept: this is a cellular concept related to calcium desensitisation of the myofibrils. The necrotic myocardium remains akinetic even though the artery responsible has been reopened [31,32].

h. Concept of hibernation: this is a vascular concept. It involves contractile abnormalities (akinesia) related to chronic hypoperfusion which is dependent on residual coronary stenosis, the removal of which (bypass or angioplasty) will significantly improve the extent and degree of this akinesia [33].

5. Healing: begins about a week after the onset of ischaemia, collagen secreted by fibroblasts replaces the inflammatory infiltrate, forming the fibrous scar of the infarct.

Clinical signs suggestive of ACS

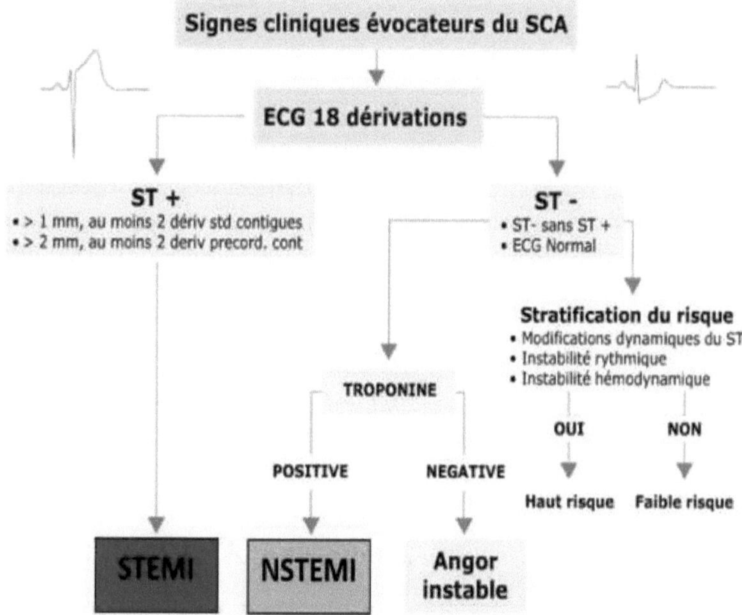

Figure 7: Classification of ACS and diagnostic approach [41].

D. Nosology of ACS :

Atherosclerosis, the leading cause of death worldwide, is the most frequent cause of MI (95%) [10].

Acute MI is systemic ischaemic myocyte necrosis secondary to acute thrombotic coronary occlusionê.

Post-mortem examinations at the end of the 19th century demonstrated a relationship between thrombotic occlusion of a coronary artery and myocardial infarction (MI) [34].

However, it was not until the beginning of the 20th century that the first clinical

descriptions of thrombus formation in a coronary artery appeared [35, 36].
The theory of plaque rupture-thrombosis is one of the essential mechanisms in the progression and acute complications of atherosclerosis. It was in 1912 that J.B. Herrick first described the association between acute myocardial infarction and coronary thrombotic occlusion [37].
In 1940, J. Duguid, a pathologist in Cardiff, revived the idea that thrombosis could be the driving force behind atherosclerosis. In 1958, Jean Lenègre published "Heart disease caused by atherosclerosis", an idea that was revived in the early 1980s when the first coronary angiographies were carried out in the acute phase of MI, showing coronary thrombosis, and then in the early 1990s with endocoronary ultrasound, showing thrombus in vivo in its atherosclerotic bed.
A general, worldwide definition of infarction first occurred in the 1950s-1970s, when WHO working groups defined MI as the combination of two of three characteristics: angina-like chest pain lasting more than twenty minutes, enzyme elevation and the appearance of Q waves of necrosis on the ECG [38, 39].
The recent development of new imaging techniques for detecting necrosis (less than 1 g) and, above all, the appearance of new, highly sensitive markers of myocyte lysis (troponins I or T) have revolutionised this approach to myocardial infarction. Indeed, since Since 2000, a consensual definition based on ECG has been established by the European Society of Cardiology and the American College of Cardiology [40], distinguishing between acute coronary syndromes with persistent STST elevation (STEMI) requiring emergency coronary decontraction and non-ST elevation acute coronary syndromes (NSTEMI) [41] where a second distinction, this time biological, is necessary to divide the group in two according to whether or not troponin is elevated, thus distinguishing unstable angina with no troponin elevation and non-ST elevation myocardial infarction (NSTEMI) : with troponin elevation.
In addition to the concept of ACS, myocardial infarction can be classified into several types according to pathological, clinical and prognostic characteristics.
The principle was refined by the Global MI Task Force, leading to the consensus document on the universal definition of myocardial infarction in 2007, introducing a new classification system for myocardial infarction with 5 sub-categories [42].
This document, approved by the ESC, the ACC, the American Heart Association (AHA) and the World Heart Federation (WHF), has been adopted by the WHO.
Through this document, the European Society of Cardiology (ESC) and the American College of Cardiology (ACC) collaborated to redefine MI through a biochemical and clinical approach, benefiting from scientific advances in cardiac biomarker detection. It was thus indicated that all myocardial lesions detected by abnormal biomarkers in the context of acute myocardial ischaemiaê should be classified as MI [43] and that the term myocardial lesion should be used when there is evidence of elevated cardiac troponin (cTn) values, with at least 1 value above the 99th percentile of the upper reference limit. Then the pattern of increasing and/or decreasing cTn values is necessary for it to be considered acute.
The most recent definition of myocardial infarction was established in a consensus document on the 4ème universal ESC/ACC/AHA/WHF definition published in 2018 [44]; the criteria for establishing the diagnosis for each category were defined as follows:

1. Type 1 myocardial infarction: MDI caused by atheromatous coronary pathology, generally provoked by the rupture of atheromatous plaque (rupture or erosion), is known as type 1 infarction.

Criteria for identifying type 1 infarction: detection of a rise and/or fall in cTn values with at least one value above the 99th percentile of the upper reference value and with at least 1 of the following :

- Symptoms of acute myocardial ischaemiaê
- New ischaemic ECG changes
- The development of pathological Q waves
- Imaging evidence of new loss of viable myocardium or new wall motion abnormality compatible with ischaemic aetiology.
- Identification of a coronary thrombus by angiography, including imaging or autopsy.

It is essential to integrate ECG findings to classify type 1 myocardial infarction as STEMI or NSTEMI in order to establish appropriate treatment in accordance with current guidelines [44].

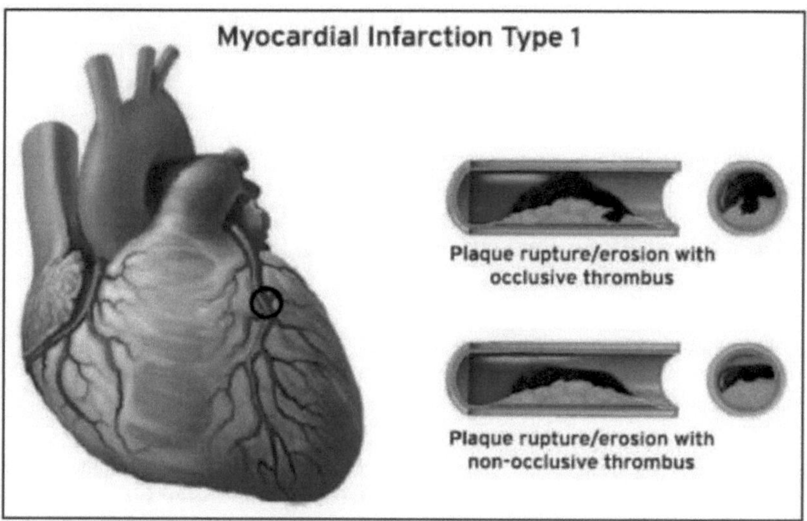

Figure 8: Type 1 myocardial infarction

2. Type 2 myocardial infarction :

The pathophysiological mechanism leading to ischaemic damage to the myocardium as a result of an imbalance between oxygen supply and demand has been classified as type 2 MI. Coronary atherosclerosis is a common finding in patients with type 2 MI. The incidence of ST elevation in type 2 MI has been shown to range from 3% to 24% [45].

Myocardial Infarction Type 2

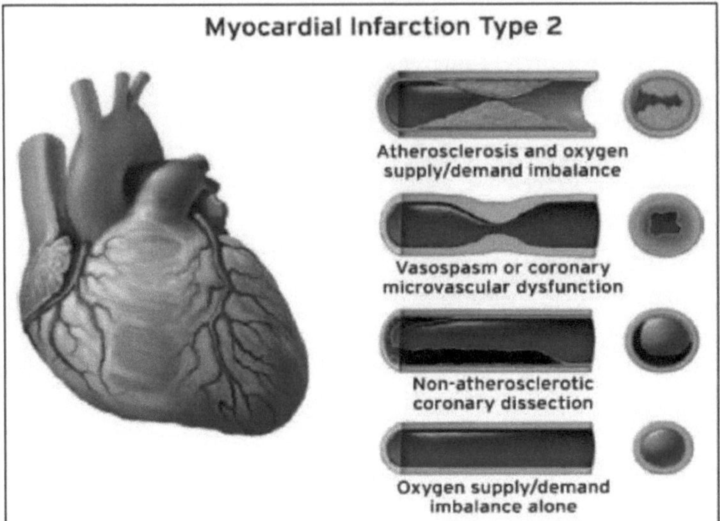

Figure 9: Type 2 myocardial infarction

Criteria for identifying type 2 infarction: detection of a rise and/or fall in cTn values, with at least one value above the 99th percentile of the upper reference value, and evidence of an imbalance between myocardial oxygen demand and supply, unrelated to acute coronary atherothrombosisê requiring at least 1 of the following:

- Symptoms of acute myocardial ischaemiaê
- New ischaemic ECG changes
- Development of pathological Q waves ;
- Imaging evidence of new loss of viable myocardium or new wall motion abnormality compatible with ischaemic aetiology.

3. Type 3 myocardial infarction :

Patients with cardiac death, with symptoms suggestive of myocardial ischaemia and new ischaemic ECG changes or ventricular fibrillation, but who died before blood samples for biomarker testing could be obtained, or before increases in cardiac biomarkers could be identified, or before MI was detected by autopsy.

4. Other types of myocardial infarction :

- **Type 4:** myocardial infarction associated with percutaneous coronary intervention (angioplasty) with 3 sub-categories: type 4a, type 4b, type 4c
- **Type 5:** myocardial infarction associated with coronary artery bypass grafting.

B. Cardiovascular risk factors:

1-Age:

With age, atherosclerosis lesions chronologically affect the aortic, coronary and then carotid arteries. Fifty for men and sixty for women is the age at which the risk of an acute coronary event begins to increase. However, the current trend is for MI to affect younger and younger people.

2-Sex:

It has been established that men are clearly more exposed to the risk of CVD. Out of 100 heart attacks, only 20 occur in women. This protection is linked to the beneficial influence of natural restrogens on the lipid profile, insulin sensitivity and blood pressure. This protection disappears 10 to 15 years after the menopause and explains the later age of onset of complications in women.

3-Coronary heredity:

A family history of CVD in first-degree relatives is a cardiovascular risk factor, especially if the events occurred at an early age in the family (father <55 and mother <65) [46].

4- Smoking :

It is one of the most serious threats to **cardiovascular morbidity and mortality** and to global public health [47]. Whether active or passive, smoking potentiates other DRFs and affects all phases of atherosclerosis, from endothelial dysfunction to acute clinical events with a thrombotic effect [48]. Smoking increases inflammation, thrombosis and oxidation of low-density lipoproteins. Smoking increases the degradation of nitric oxide, which has a cytotoxic effect on the myocardium. After **smoking cessation**, the increase in relative risk disappears within 3 years and reduces the risk of recurrence in coronary heart disease by 50% [49].

5- Diet :

It is the most important behavioural risk factor, after smoking, in determining coronary risk. The atherogenic effect of diet is based in part on the consumption of saturated fatty acids, which increase LDL cholesterol. The WHO also recommends reducing **sugar** intake **to around 25 grams (6 teaspoons)** [50] **and salt to less than 5 grams a day** to lower blood pressure and the risk of CVD, stroke and MI [51].

6- Sedentary lifestyle :

A sedentary lifestyle (physical inactivity) is a lack of minimal physical activity corresponding to exercise that causes the onset of breathlessness and an acceleration of the heart rate (walking, cycling) for 30 minutes or 3 times 10 minutes, every day(). A sedentary lifestyle is considered to be the fourth leading risk factor for death worldwide (6%). It is estimated to be the main cause of 27% of cases of diabetes and around 30% of cases of ischaemic heart disease [52, 53].

7- Obesity and overweight:

Obesity was recognised as a disease in 1997 by the WHO. It defines overweight and obesity **as "an abnormal or excessive accumulation of body fat that can damage health"** [54]. The measurement indicator used is **the body mass index** (BMI= weight(Kg)/height^2 (m)). **Waist circumference** is a second anthropometric indicator that reflects abdominal adiposity and insulin resistance; it is better correlated with the risk of MI and sudden death than BMI.

The risk of ischaemic heart disease is increased by 25% in overweight subjects and by 64% in obese subjects [55].

8- Stress and the role of sleep :

When **stress** is prolonged and becomes uncontrollable, it can lead to a syndrome of exhaustion, combining strong irritability, demoralisation and lack of energy. It **is** accompanied by **hypercoagulability**, as evidenced by **an increase in plasma levels of**

fibrinogen and factor VII and a reduction in fibrinolysis [56]. According to the Fédération Française de Cardiologie (French Federation of Cardiology), **33% of heart attacks have stress as the main FDR, along with** smoking, dyslipidemia, type II diabetes and hypertension [57].

Lack of sleep, and in some cases **excessive sleep**, is associated with a number of health problems, including CVD, type 2 diabetes, hypertension, obesity and general mortality. A study published in 2019 in the Journal of the American College of Cardiology confirms that insufficient sleep is harmful to cardiovascular health. This study shows that people who slept **less than 6 hours a night had a 27% higher risk of suffering from atherosclerosis** [58].

9- Psychotropic substances :

Cannabis is the most commonly used drug in the world, mainly for its euphoric and hallucinogenic effects. **The cardiovascular effects of cannabis** involve activation of the sympathetic system, reduced activity of the parasympathetic system and action on cannabinoid receptors [59]. Cannabis increases the risk of MI by a factor of 4.8 in the 60 minutes following inhalation of a cigarette, probably due to an increase in myocardial oxygen demand [60].

10- Dyslipidemia:

Dyslipidaemia covers all disorders of lipid metabolism. We can describe: pure hypercholesterolaemia; pure hypertriglyceridaemia; mixed hyperlipidaemia: a combination of the two.

- HDL hypoemia may be associated with one or other of these categories.

The Agence Française de Sécurité Sanitaire des Produits de Santé (2005) considers a lipid profile to be normal if: - LDL-C < 1.6 g/L; HDL-C > 0.40 g/L in men and > 0.50 g/L in women; - TG < 1.5 g/L [61].

Dyslipidaemia is a major cardiovascular risk factor, often undiagnosed [62, 63]. Hypercholesterolaemia promotes endothelial dysfunction. At lipid level,
oxidation of LDL increases the expression of adhesion molecules on the surface cells of the endothelium, whereas HDL-cholesterol inhibits it. In addition, HDL promotes vasodilatation of the coronary arteries.

11- Hypertension :

A continuous and prolonged rise in pressure encourages damage to the intima and the development of atherosclerotic plaque [64, 65]. On the other hand, the increase in pressure on the vascular walls acts by causing rupture of the atherosclerotic plaque, which is the cause of acute complications (MI and stroke). Studies of ischaemic heart disease show that the incidence of hypertension exceeds 44% [66] and rises to over 75% in patients aged over 65 [67, 68]. High blood pressure has been shown to increase the risk of a cardiac event by a factor of 6 [69].

12- Diabetes :

The WHO defines **it** as fasting venous glucose **>1.26 g/L** or random venous glucose **>2 g/L** or venous glucose **>2 g/L** two hours after ingestion of 75 g of glucose. An HBA1C level **>6.5%** has been added to this definition [70, 71].

Diabetes is both a chronic disease and a major FRDCV. Type II diabetes leads to smooth muscle cell dysfunction, prothrombotic abnormalities and inflammation. **Diabetic patients are at greater risk of cardiovascular mortality and morbidity.** Among patients hospitalised for ACS, 30-40% have diabetes and 25-36% have impaired fasting glucose or impaired glucose tolerance [72]. There is a highly significant linear correlation between HbA1c and the risk of myocardial infarction [73]. Furthermore, diabetes is a poor prognostic factor after ACS [74]. Diabetes increases the risk of coronary heart disease by a factor of 2 to 4 [75]. Diabetes **is an independent predictor of mortality and recurrent infarction.**

13- Metabolic syndrome :

The metabolic syndrome (MetS) is a group of symptoms resulting from metabolic disorders. It is increasingly regarded as **a major public health problem** [76, 77].

According to the IDF, a person has DM [78] when they **have abdominal obesity** (a waist circumference greater than 94 cm in men and 80 in women) and at least two of the following factors: - TG > 1.5 g/L

- **HDL< 0.4 g/L** in men and O.5 g/L in women.
- **BP >** or equal to 130/85 mmHg or **known and treated hypertension.**
- **venous glycaemia** = or > 5.6 mmol/L or presence **of type 2 diabetes.**

In the Finnish Isomaa study [79], **the presence of SM increased the relative risk of coronary events by a factor of 3 (p<0.001) and increased cardiovascular mortality (12.0% compared with 2.2% in the absence of SM).**

14- Chronic renal failure:

The prevalence of **chronic kidney disease** continues to rise worldwide due to the ageing of the population and the increasing incidence of type 2 diabetes mellitus and hypertension [80, 82]. It is also **an independent risk factor for cardiovascular mortality and morbidity** [83, 84].

Even moderate CKD increases **cardiovascular** risk by a factor of 3 to 5.

C. Epidemiology of ACS :

The epidemiology of ACS is of major interest for public health. Ischaemic heart disease is currently the leading cause of death and years of life lost worldwide [85].

In 2015, around 17.7 million deaths (45% of all NCD-related deaths) worldwide were caused by cardiovascular disease (31% of total global mortality).

More than three quarters of CVD-related deaths occur in **low- and middle-income countries** (2015 figures) **[86].**

Ischaemic heart disease **and stroke** were responsible for a total of 15.2 million deaths in 2016. They have remained **the leading causes of death worldwide over the last 15 years** (see Figure 10) [87] (90).

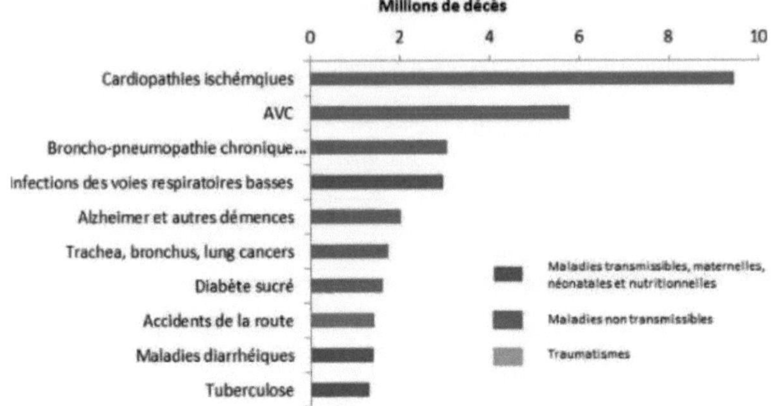

Figure 10: The 10 leading causes of death worldwide, 2016[87].

There is a downward trend in the mortality and incidence of CVD in developed countries as a result of effective primary prevention and management strategies [88, 89].

Although age-specific cardiovascular mortality rates have improved significantly in low- and middle-income countries, they remain much higher than in high-income countries.

In France, there are 100,000 cases of ST+ ACS per year. The annual incidence of STEMI hospitalised before 24 hours is close to 35,000 cases. 30-day mortality has fallen from 13.7% in 1995 to 4.4% in 2010. This reduction is probably related to changes in management: shorter time to call the EMS (from 120 to 74 minutes) with earlier management; an increase in reperfusion (from 49.4% to 74.7%) [90].

In Algeria, a central system for recording causes of death was set up by the National Institute of Public Health (INSP) in 1995. Analysis of the causes of death in 2002 showed that deaths due to NCDs ranked first, with a rate of 58.5%. In the NCD group, cardiovascular pathology was the leading cause of death, accounting for 44.5% [91].

According to the TAHINA study in 2002, 38.7% of deaths in Algeria occur after the age of 70. Cardiovascular disease is the leading cause of death. From 26.2% of the general population, the proportion of deaths has risen to 40% in the over-45s.

In 2013, 13,621 deaths secondary to diseases of the circulatory system were recorded and treated by the INSP, giving a proportion of 21.6% (first place) of all deaths occurring nationally [92]. The study of the annual incidence of ST+ ACS in 2013 in western Algeria, calculated in the adult population aged over 20 years, was 34/100,000 inhabitants, and the RECORD register of ACS patients hospitalised at the Hussein-Dey University Hospital in 2013 found a hospital mortality rate of 9.1% for ST+ ACS (and 2% for ST- ACS). The "RECORd" register in its 2014 analysis shows that 842 patients with acute coronary syndrome were hospitalised in the department, 505 cases of ACS with ST-segment elevation and 337 without ST-segment elevation (ratio of 1.49). A national estimate based on these data suggests that there are 24,000 ACS per year in Algeria (15,000 ST+ ACS per year and 9,000 ST- ACS per year) [93].

Between February 2014 and July 2015, 467 patients with ST-segment elevation acute coronary syndrome were hospitalised at CHU Hussein Dey with a sex ratio of 4.36. The mean age was 60.18 ± 12.62 years (extremes 23 and 93 years). The mean age of the women was 66.56 years, while that of the men was 58.71 years. The 48-hour hospital case fatality rate was 3.64% [94]. In 2016, according to WHO data, NCDs (cardiovascular diseases (CVD), cancers, chronic respiratory diseases and diabetes) accounted for 76% of all deaths in Algeria, i.e. 144,000 deaths. Proportional mortality from CVD was 36%. The probability of dying between the ages of 30 and 70 from one of the 4 main NCDs is 14% (15% for men and 13% for women) [92].

D. Diagnosis of STEMI :

1 Diagnostic strategy :

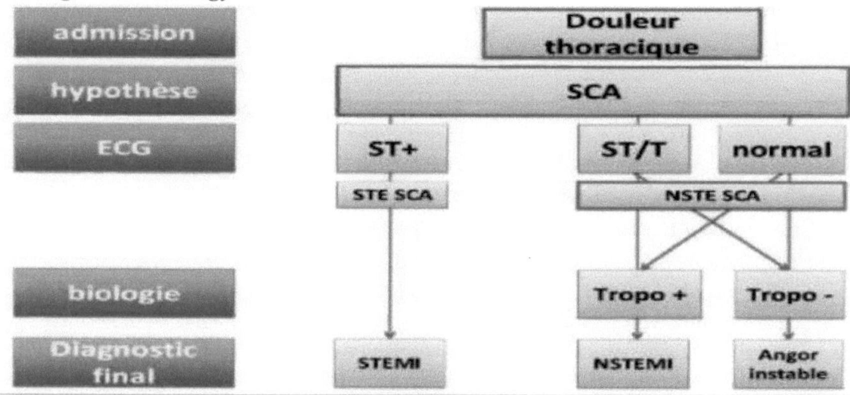

Figure11: Classification of Sca [41].

ACS is a clinical entity which groups together myocardial ischaemia in their old terminologies: unstable angina, non-Q wave myocardial infarction and Q wave myocardial infarction. In the absence of troponin elevation, the term unstable angina is used [95].

ACS is a pathology that is observed throughout the world [96]. Its prevalence was 5-20% in Europe and 13.5% in Africa [97, 98].

It constitutes a diagnostic and therapeutic emergency which may be life-threatening in the short term. Mortality is considerable, at 16-20% in the first few hours [99].

When a patient complains of chest pain, the primary objective is to detect the presence of signs of seriousness that would require urgent specific treatment, and to clarify the aetiological diagnosis.

b. Search for signs of seriousness :

These are :

- Respiratory signs: dyspnoea with elements of severity, cyanosis, sweating, draughts supra-sternal, supra-clavicular, thoraco-abdominal rocking, inability to speak, arterial oxygen saturation below 90%;
- Haemodynamic signs: tachycardia, rhythm disorders, hypotension arterial hypertension, signs of peripheral hypoperfusion (mottling, longer skin recolouration time) or even cardiocirculatory arrest;
- Neurological signs: confusion, agitation, disturbed consciousness, coma.

In the presence of these signs, symptomatic treatment is urgently required before diagnosis.

c. Search for a diagnosis requiring specific and urgent treatment:

Four diagnoses need to be considered:

- Aortic dissection ;
- Pulmonary embolism with signs of severity ;
- Tamponade ;
- ST-segment elevation ACS.

They require urgent treatment, the implementation strategy for which includes referral to the appropriate care structure for the patient's condition. They account for the bulk of initial treatment.

d. Diagnostic approach :

In the presence of chest pain, the diagnostic strategy relies primarily on the clinic and the electrocardiogram. Recently, however, the possibility of carrying out biological assays (enzymology, natriuretic peptide B, d-dimer, etc.) or ultrasound scans as early as the pre-hospital care phase, is helping to obtain a precise diagnosis that will enable rapid therapeutic decisions to be taken.

e. Diagnosis of acute coronary syndrome :

The diagnosis of ACS is based on research [100]:

- cardiovascular risk factors: family history, male sex, advanced age, smoking, diabetes, high blood pressure, high cholesterol, etc.
- the circumstances in which the pain occurs, typically at rest in the morning, but sometimes preceded by other painful episodes. It may also be a case of pain appearing on exertion, or of a previously known angina that has become unstable, i.e. with more frequent attacks and/or triggered by lesser exertion and/or which is less responsive to nitroxide;
- characteristics of the pain, typically retrosternal, constrictive, with cervical, scapular or brachial radiations. By definition, the pain persists for more than 20 minutes;
- the clinical examination, which is typically normal, is mainly used to rule out another diagnosis: normal and symmetrical blood pressure, normal cardiac and pulmonary auscultation, perceptible peripheral pulses, supple and painless calves, etc. The clinical examination can also be used to look for complications such as tachycardia, acute pulmonary redema or cardiogenic shock, all of which indicate poor clinical tolerance;
- the electrocardiogram, which plays a major role in cases of suspected ACS [101]. Firstly, because in most cases it will confirm coronary damage, and secondly, because it enables patients to be classified according to whether or not they have ST-segment elevation. This classification gives rise to different therapeutic strategies. In a setting of suggestive chest pain, the presence of ST-segment elevation (in at least two leads of the same coronary territory) confirms the diagnosis of infarction, raises suspicion as to the coronary vascular territory affected and indicates the urgent need to implement a reperfusion strategy. It should be remembered that certain situations, such as right bundle branch block, left bundle branch block or electroentrainment (see below), do not preclude the diagnosis of ST-segment elevation ACS. The presence of a mirror or even a Q wave increases the probability of the diagnosis.

Some signs in favour of a diagnosis of ST-segment elevation ACS have been described more recently:

- Sgarbossa signs, used to confirm the presence of an ST elevation in the presence of left bundle-branch block (or electroentrainment): loss of discordance between QRS complexes and repolarisation or the existence of a "super-ST elevation" (greater than 5 mm) in the right precordial leads [102- 104];

- The presence of "fractured" QRS, QRS presenting

low amplitude, jagged hooks resembling artefacts. These signs were associated with infarct scarring with a sensitivity, specificity and negative predictive value of 87%, 89% and 93% respectively (higher than those of the Q wave) [104];

- The presence of ST-segment sub-shift in the anterior precordial leads, which indicates a posterior infarct [105].

Other electrocardiographic abnormalities: large, positive, sharp and symmetrical T waves or, conversely, inverted T waves, signs of ischaemia; ST segment sub-shift; transient ST segment elevation or, later, the presence of a pathological Q wave (longer than 0.04 seconds and more than one third of the amplitude of the QRS complex) enable the patient to be classified as having ACS without ST segment elevation. Priority is then given to risk assessment. In this context, the diagnostic approach is often supplemented by biological tests, particularly enzyme tests (troponin). Whatever the criteria or score used (ESC criteria, TIMI score or GRACE score), risk stratification in these patients is impossible [106-108]. When the risk is high, an invasive strategy combining the administration of an anti-GPIIb/IIIa and early coronary angiography could be useful [109, 110].

a. Case history :

It is used to record the characteristics of the chest pain. It is an angina pain, lasting more than thirty minutes, it is wide, retrosternal radiating towards the left arm, sometimes towards both arms or towards the jaw or interscapular. This pain usually occurs at rest, resists sublingual nitrate derivatives, is very distressing and accompanied by sweating, and can sometimes reach a paroxysm. For the purposes of reperfusion, the time of onset of the last prolonged pain should be taken as the reference point.

However, atypical forms of pain are frequently encountered in diabetics, the elderly and women.

b. Clinical features :

ST+ ACS corresponds **to ischaemic necrosis** of a myocardial region associated with **complete and prolonged occlusion of** a coronary artery. Its clinical manifestation is prolonged (> 20 to 30 minutes) constrictive, intense retrosternal pain radiating to the jaws and neck vessels, most often occurring at rest. It is a **nitroresistant** pain, but the diagnostic test (sublingual trinitrine), useful in cases of atypical pain, is contraindicated in cases of right ventricular infarction and systolic blood pressure below 90 mmHg, just as it is not recommended in cases of inferior infarction.

The physical examination is particularly useful for looking for complicated forms (ventricular failure, mechanical complications, etc.).

c. Electrocardiography :

Various ECG abnormalities follow one another in the first few hours: subendocardial ischaemia, subepicardial injury current (Pardee wave) associated with mirror images, then Q wave of necrosis.

An ample positive and sharp T wave corresponding to sub-endocardial ischaemia can be seen during the first hour of a myocardial infarction. Unfortunately, this early sign is very fleeting. The pathognomonic electrocardiographic sign of a developing infarction is an

ST-segment elevation of at least one millimetre in the frontal leads and at least two millimetres in the precordial leads, convex upwards and encompassing the T wave, producing the classic Pardee wave. This sign, together with chest pain, is sufficient to initiate an emergency drug reperfusion strategy in the absence of contraindications. A mirror sub-shift is usual in the leads diametrically opposed to those of the infarct. The Q wave of necrosis usually appears between the fourth and sixth hour and sometimes necrosis may be reflected by a "planing" of the R waves in the same territory.

d. Troponin assay :

Troponins are the reference biomarker for infarcts caused by plaque rupture. Their significant elevation reflects myocardial or striated muscle cell catabolism. The troponin complex is made up of 3 sub-units - troponin C, troponin I, troponin T - which are located on the fine filament (actin) of the contractile apparatus of striated skeletal and cardiac muscles. These proteins regulate actin-myosin interaction during muscle contraction. The cardiac isoforms of troponins I and T (cTn I, cTn T) are only expressed in cardiac muscle. Ordinary troponins appear around the fourth hour of pain, peaking at the twelfth hour (sensitivity 100%, specificity 94%) and remain elevated for ten days in the absence of reperfusion (signs of myocardial cell necrosis, troponins are released after 4 h, peak around 14 h, elevation persists 75 to 140 h for troponin I and more than 10 days for troponin T). These elevations are proportional to the extent of the infarct. They are accelerated by reperfusion (thrombolysis or angioplasty).

With the emergence of the hypersensitive troponin (hs) assay, lower rates of myocardial damage are detectable within the first hour if a high-sensitivity test is used.

e. Echocardiography :

At this stage, echocardiography can be used to determine the location and extent of the abnormalities in segmental kinetics that characterise myocardial infarction, and also to assess global left ventricular function using the Simpson method. Compensatory hyperkinesia of the walls facing the necrotic wall can also be detected. Echocardiography will also look for a mechanical complication or thrombus in the akinetic zone.

f. Risk assessment :

The risk of complications can vary significantly depending on the clinical and biological parameters presented by patients. Various risk stratification scores have been developed from registry data. In the context of ST+ ACS, the scores are useful for assessing medium- and long-term prognosis [111].

The scores currently recommended are :

f1. TIMI risk score :

This score has prognostic value for weighing up the severity of the infarction for each patient and their risk of mortality at D30 of ST+ ACS. It remains useful in practice for identifying patients at high coronary risk and encompasses several parameters (Table 1):

- **Clinical** (age, weight, history of diabetes, hypertension or angina, infarct topography),
- **Hemodynamics** (blood pressure, heart rate and Killip grade) and the time taken for thrombolysis.

Table I: Calculation of TIMI risk for ST+ ACS from Morrow DA et al [113].

Risk factors	Number of points
Age 65-74 / > 75	**2 / 3**

Systolic blood pressure < 100	3
Heart rate > 100	2
Kilip II-IV	2
Anterior Sus-ST or LBB	1
Diabetes, hypertension or angina	1
Weight < 67 kgs	1
Time between pain and treatment > 4 hours	1
Calculation of the TIMI ST+ risk score (cumulative points)	Mortality at 30 days (% of total)
0	0.8
1	1.6
2	2.2
3	4.4
4	7.3
5	12.4
6	16.1
7	23.4
8	26.8
> 8	35.9

f2. The GRACE score :

Developed from an international registry of patients hospitalised for ACS (ST+ or non-ST+), it is based on various clinical, biological and electrocardiographic criteria. It is used to assess the risk of death or infarction within the hospital and over the medium term (6 months).

However, the complexity of calculating this score requires the use of a computer or smartphone, making it difficult to use, particularly in pre-hospital settings [112].

E. Management of ST-segment elevation ACS :

I. Current information on the subject :

STEMI is an extreme coronary emergency requiring early and appropriate treatment. Acute coronary syndrome (ACS) occurs three to four times more often in men than in women under 60. After a coronary occlusion, the speed with which ischaemia then myocardial necrosis sets in means that urgent deobstruction is required to avoid the two main complications, which are either sudden death of the patient or necrosis of the ischaemic territory.

To achieve this therapeutic objective, two types of coronary deobstruction are used to restore flow and relieve ischaemia:

- Primary angioplasty: should ideally be performed within the first 120 minutes; it is effective in more than 90% of cases [289, 290]. However, it requires the availability of a catheterisation room and a cardiac intensive care unit (USIC) [284, 285];
- or thrombolysis, which is easy to perform anywhere during the first 12 hours, but is only effective in 60% of cases [286, 287]. The prognosis for thrombolysis is further improved if it is started early, especially in the pre-hospital setting [288-291]; it is indicated regardless of the age, sex or condition of the high-risk patient [292] or even in shock [293].

II. Strategy:

All patients presenting with STEMI should be managed according to their level of ischaemic and haemorrhagic risk. Emergency management and transport should be organised with the aim of carrying out a **reperfusion** procedure **within the first few hours for** all patients with ST+ ACS **less than 12 hours old.** The objectives of therapeutic management are to reduce mortality, prevent complications and relieve pain.

In the absence of urgent reperfusion, complications may be **haemodynamic** (acute heart failure, cardiogenic shock), **rhythmic** (sudden death from ventricular fibrillation, ventricular tachycardia or asystole), **mechanical** (wall rupture leading to ventricular septal defect or tamponade, mitral pillar rupture causing massive mitral insufficiency) or **thromboembolic**.

3. Emergency drug treatment combines :

- morphine analgesics ;
- oxygen therapy (if SaO2 < 95%);
- double platelet anti-aggregation (according to the recommendations of the European Society of Cardiology 2017): in the absence of specific contraindication, Ticagrelor or Prasugrel are recommended as first-line treatment, combined with Aspirin. In the event of thrombolysis, only clopidogrel is recommended and a relay with ticagrelor or prasugrel may be proposed from day 3 [114];
- an anticoagulant: unfractionated heparin or enoxaparin.

The coronary reperfusion strategy depends on the **time taken for the infarction to progress and** the **time required to perform the angioplasty.**

4. Primary angioplasty: is the recommended reperfusion therapy of choice in STEMI, if performed by an experienced team within **120 minutes** of the initial medical contact. In most cases, the procedure is completed by the insertion of a metal **stent** (see Figures 12 and 13).

Randomised clinical trials comparing primary angioplasty performed on time, in experienced centres with a high volume of interventional procedures, compared with hospital fibrinolysis, have repeatedly shown that
that the first reperfusion therapy was superior to the second [117-120].

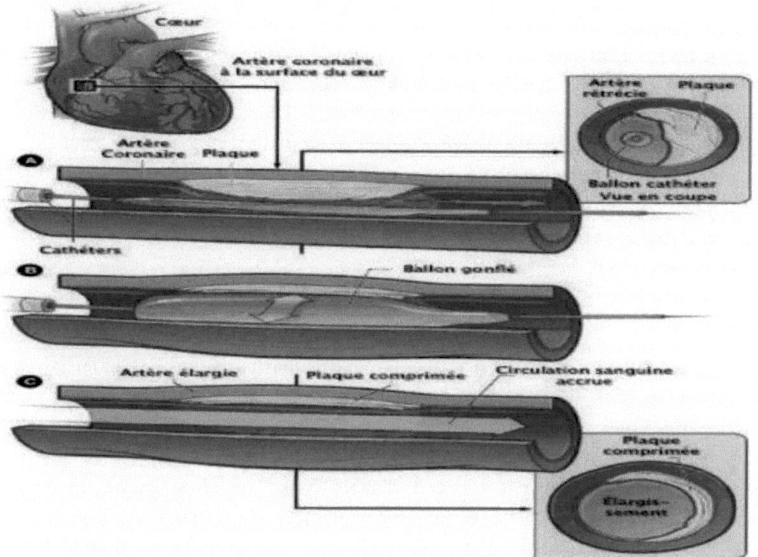

Figure 12: Angioplasty without stenting [115].

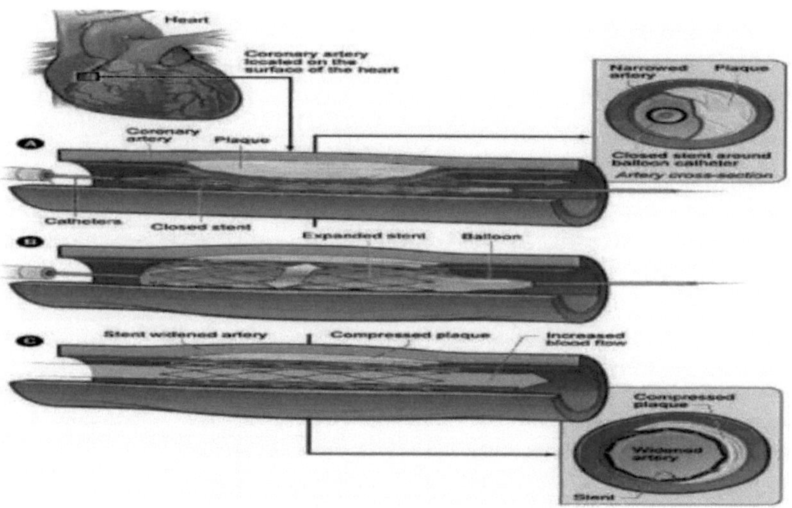

Figure 13: Angioplasty with stent placement [116].

5. **Intravenous fibrinolysis** is administered if the time required to perform angioplasty after a qualifying ECG cannot be met. The benefits of fibrinolysis are well established compared with placebo. It prevents approximately 30 deaths per 1000 patients treated within 6 hours of the onset of symptoms. If possible, fibrinolysis should be started pre-hospital. The later the patient presents [121-123], the more consideration should be given to transferring the patient for PCI, as the efficacy and clinical benefit of fibrinolysis diminish as the time to onset of symptoms increases.

6. **Pre-hospital fibrinolysis:** In a meta-analysis of six randomised trials, pre-

hospital fibrinolysis reduced early mortality by 17% compared with hospital fibrinolysis, with a time saving of approximately 60 minutes [124], particularly when administered within the first 2 hours after the onset of symptoms (44% reduction in mortality for the first 2 hours compared with 22% after that) [125]. These data support the pre-hospital initiation of fibrinolytic treatment when a reperfusion strategy is indicated [126-128]. If the medical or paramedical staff on site are trained and able to analyse the ECG on site or transmit the ECG to the hospital for interpretation, it is recommended that fibrinolytic therapy be initiated pre-hospital. The aim is to start fibrinolytic therapy within 10 minutes of the diagnosis of STEMI [129].

7. Angiography and percutaneous coronary intervention after fibrinolysis :

After initiation of lytic therapy, it is recommended that patients be transferred to a PCI centre. If fibrinolysis fails, or if there is evidence of reocclusion or reinfarction with recurrence of ST elevation, immediate angiography and salvage PCI are indicated [130].

Even if fibrinolysis is successful (ST segment resolution >50% at 60-90 minutes, arrhythmia typical of reperfusion and disappearance of chest pain), a strategy of systematic early angiography is recommended if there are no contraindications. Several randomised trials [131-135] and meta-analyses [136,137] have shown that early systematic angiography followed by PCI (if necessary) after fibrinolysis reduces the rates of reinfarction and recurrent ischaemia. Thus, early angiography followed by PCI if necessary is also the recommended standard of care after successful fibrinolysis.

A crucial issue is the optimal time between successful fibrinolysis and PCI. The time varies considerably from one trial to another, ranging from a median of 1.3 h in the CAPITAL AMI trial [134] to 17 h in the GRACIA1 [135] and STREAM [138] trials.

In an analysis of six randomised trials, very early angiography (<2 h) after fibrinolysis was not associated with an increased risk of death/reinfarction at 30 days or major bleeding, and a shorter time from symptom onset to angiography (<4 h) was associated with a reduced risk of death/reinfarction at 30 days and 1 year and recurrent ischaemia at 30 days [139]. On the basis of this analysis, and of trials in which the median time between the start of thrombolysis and angiography was 2-17 h [138, 131, 132, 133], a time window of 2-24 h after successful fibrinolysis is recommended [129].

Table II: Target times according to the 2017 ESC recommendations

Intervals	Target times
Maximum time between first medical contact, ECG and diagnosis	< 10min
Maximum expected time between STEMI diagnosis and primary angioplasty (passage of the wire) to choose the strategy of primary angioplasty rather than fibrinolysis (if this target time cannot be met, consider fibrinolysis)	< 120 minutes
Maximum time from STEMI diagnosis to wire passage in patients presenting to hospitals performing primary angioplasty	< 60min
Maximum time between diagnosis of STEMI and passage of the wire in transferred patients	< 90min
Maximum time from STEMI diagnosis to start of fibrinolysis bolus or infusion in patients unable to meet target times for primary angioplasty	< 10min
Time between the start of fibrinolysis and assessment of its	60-90min

effectiveness (success or failure)	
Time between the start of fibrinolysis and angiography (if fibrinolysis is successful)	2-24 hours

8. Concomitant medical treatment

In addition to early reperfusion therapy, a variety of drugs are recommended for STEMI patients by international guidelines [129, 140]. The choice of anticoagulant and antiplatelet therapy depends on the reperfusion strategy and the patient's ischaemic and haemorrhagic risks.

a. Anticoagulant treatment.

Anticoagulation treatment aims to inhibit the formation and activity of thrombin, which plays an important role in the pathophysiology of STEMI. Several options are available to provide rapid and effective anticoagulation for STEMI patients treated with primary PCI. Antithrombin III (AT3) is a peptide that inhibits several of the activated coagulation factors. Unfractionated heparin binds to AT3 and increases its activity, and also has a certain inhibitory effect on coagulation factors IXa, XIa and XIIa. Enoxaparin is a low molecular weight heparin and also binds to AT3. In primary PCI, these agents should be administered intravenously [129, 140]. Whereas the dose of unfractionated heparin must be adjusted on the basis of an activated partial thromboplastin time measurement, enoxaparin requires no monitoring and is administered as a single bolus.

UFH dosage should follow the standard recommendation for PCI (i.e. initial bolus of 70-100 U/kg). An I.V. bolus of enoxaparin 0.5mg/kg was compared to UFH in the ATOLL randomised trial, including 910 STEMI patients [141], I.V. enoxaparin was superior to UFH in reducing ischaemic events, mortality and major bleeding [142].

In a meta-analysis of 23 PCI trials, enoxaparin was associated with a significant reduction in death and major bleeding compared with UFH. This effect was particularly significant in the context of primary PCI [143]. On the basis of these considerations, enoxaparin should be considered in STEMI.

Bivalirudin is a direct thrombin inhibitor and can be used in patients at high risk of bleeding or with heparin-induced thrombocytopenia.

Unless there is a clear indication for continuing anticoagulant treatment (atrial fibrillation, mechanical valve prosthesis or intraventricular thrombus, among others), curative anticoagulant treatment is not generally indicated after PCI. Prophylactic doses for the prevention of venous thromboembolism may be indicated in patients on prolonged bed rest [129].

Patients receiving fibrinolysis should also receive anticoagulant therapy until PCI is performed (if feasible) or for the duration of the hospital stay (but not exceeding 8 days). Enoxaparin is preferable to unfractionated heparin; in patients treated with streptokinase, fondaparinux should only be used with streptokinase [129].

In the ATLAS ACS 2-TIMI 52 trial, low-dose rivaroxaban (2.5 mg twice daily), in addition to aspirin and clopidogrel, reduced cardiovascular death, MI, stroke, and all-cause mortality over a mean follow-up of 13 months [144]. Stent thrombosis was reduced by one third. However, this was associated with a three-fold increase in the risk of major bleeding, particularly intracranial haemorrhage [144].

According to the same study, in selected patients at low risk of bleeding, rivaroxaban 2.5 mg can be considered in patients on dual antiplatelet therapy after STEMI.

b. Antiplatelet treatment.

Antiplatelet agents aim to prevent platelets from forming a thrombus and are therefore essential for the treatment of patients during and after STEMI. The standard of care for antiplatelet therapy in STEMI is oral dual antiplatelet therapy combining lifelong aspirin and an oral inhibitor of the P2Y12 receptor (P2Y12; the predominant receptor in ADP-stimulated prolonged platelet aggregation) [145], which, as a general rule, should be used for 12 months [146].

Detailed guidelines have been published on the optimal duration of dual antiplatelet therapy [147, 148].

In all patients, aspirin should be given as a loading dose as soon as possible after diagnosis, and treatment should be maintained at a low dose (75-100mg).

Aspirin can be administered orally, or as an I.V. loading dose to ensure complete inhibition of thromboxane A2, on which platelet aggregation depends.

The loading dose of oral aspirin should preferably be 150 to 300 mg. There are few clinical data on the optimal i.v. dose. A randomised trial showed that a single dose of 250 or 500mg of acetylsalicylic acid intravenously compared with 300mg orally was associated with more rapid and complete inhibition of thromboxane generation and platelet aggregation at 5 minutes, with comparable rates of bleeding complications [149].

There is limited evidence on the best timing to administer P2Y12 inhibitor, the ATLANTIC (Administration of Ticagrelor in the Cath Lab or in the Ambulance for New ST Elevation Myocardial Infarction to Open the Coronary Artery) trial [150] is the only randomised study testing the safety and efficacy of different timing of P2Y12 inhibitor initiation in STEMI. Early initiation of P2Y12 inhibitor therapy during patient transport to a primary PCI centre is common practice in Europe and is consistent with pharmacokinetic data.

In addition, early treatment with high-dose clopidogrel has been shown to be superior to treatment during catheterisation in observational studies and in a randomised trial [151, 152, 153].

Overall, the data suggest that early administration may be preferable for good efficacy.

There are three options for oral P2Y12 inhibitors.The preferred P2Y12 inhibitors are prasugrel (60mg loading dose and 10mg maintenance dose once daily po) or ticagrelor (180mg loading dose po and 90mg maintenance dose twice daily). These drugs have a faster onset of action, are more potent and are superior to clopidogrel in terms of clinical results [154, 155].

Thus, current guidelines [129, 140] recommend potent oral P2Y12 inhibitors (ticagrelor or prasugrel) over clopidogrel, given the advantages of these agents over clopidogrel in the results of large trials [154, 155]. Beneficial effects on reducing ischaemic events, albeit with an increased risk of bleeding, have been documented [154, 155]. Prasugrel is contraindicated in patients with a history of stroke or transient ischaemic attack and is not recommended in patients aged >75 years. In patients weighing <60 kg, the maintenance dose should be reduced.

However, clopidogrel remains the drug of choice in patients at high risk of bleeding, particularly patients requiring lifelong oral anticoagulation, or if the new molecules are unavailable, contraindicated or poorly tolerated. Clopidogrel has not been evaluated in large studies in primary PCI, but a higher regimen of 600 mg loading dose and 150 mg maintenance dose during the first week was superior to the 300/75 mg regimen in patients undergoing PCI in the CURRENT-OASIS 7 study (Clopidogrel and aspirin Optimal Dose usage to reduce recurrent events- Seventh organization to assess strategies in ischaemic syndromes), [156] and the use of high loading doses of clopidogrel resulted in more rapid inhibition of the adenosine diphosphate receptor. All P2Y12 inhibitors should be used with caution in patients at high risk of bleeding or with significant anaemia.

Intravenous antiplatelet agents include GPIIb/IIIa receptor blockers and cangrélor, a P2Y12 inhibitor. Currently, GPIIb/IIIa receptor blockers are mainly used in the catheterisation suite to manage complications arising during PCI, such as large thrombus, slow flow (delayed progression of injected contrast into the coronary tree) or absence of reflux. Cangrelor has a rapid onset and onset of action and may reduce periprocedural ischaemic complications, although it is associated with an increased risk of bleeding compared with clopidogrel [157]. Cangrelor may be considered in patients not pre-treated with oral P2Y12 inhibitors at the time of PCI or in those considered unable to take oral inhibitors.

In patients receiving fibrinolysis, dual anti-platelet therapy should be limited to lifelong aspirin and clopidogrel for 12 months in the event of subsequent PCI, and 6 months in the event of fibrinolysis without subsequent PCI. Prasugrel and ticagrelor have not been studied as adjuvants to fibrinolysis, as their safety (i.e. bleeding complications) is not well established in this setting [129].

While there are no studies on the optimal duration of dual antiplatelet therapy in patients at high risk of haemorrhage, multiple studies have shown that reducing the duration of treatment to 6 months, compared with 12 months or more, reduces the risk of major haemorrhagic complications, with no implications for ischaemic events [158, 159].

c. Systematic secondary prevention treatments

In addition to anticoagulant and antiplatelet therapy, several pharmacological therapies should be considered and established prior to hospital discharge in STEMI patients [129].

d. Early administration of intravenous ß-blockers :

In patients undergoing fibrinolysis, early treatment with IV beta-blockers reduces the incidence of malignant ventricular arrhythmias, although there is no clear evidence of long-term clinical benefit [160, 161, 162].

In patients undergoing primary PCI, the METOCARD-CNIC trial showed that very early administration of IV metoprolol (15 mg) at the time of diagnosis in patients with previous STEMI and no evidence of heart failure and a SBP >120 mmHg was associated with a reduction in infarct size measured by cardiac MRI at 5-7 days, and a higher LVEF at 6 months compared with control treatment [163, 164]. All patients with no contraindications received oral metoprolol within 24 hours. The incidence of major cardiovascular events (cardiac death, admission for heart failure, reinfarction or malignant ventricular arrhythmias) at 2 years was 10.8% in the groups receiving IV metoprolol compared with 18.3% in the control group [164]. Treatment with metoprolol was associated with a significant reduction in the incidence and extent of microvascular obstruction [165].

On the basis of currently available data, early administration of IV beta-blockers on admission, followed by oral beta-blockers, should be considered in haemodynamically stable patients undergoing primary PCI.

e. Medium- and long-term administration of oral beta-blockers :

The widespread use of ß-blockers after MI is based on several large clinical trials [166, 167, 168], and one review supports their use, showing a reduction in mortality and morbidity [169]. Although most of these reference data come from trials conducted before the era of reperfusion. A recent multicentre registry that enrolled 7057 consecutive patients with MI showed a benefit in terms of reduced mortality over a median follow-up of 2.1 years associated with the prescription of beta-blockers at discharge, but no relationship between dose and outcome could be identified [170].

Beta blockers after MI also improve survival [171, 172]. However, in a recent observational study, treatment with ß-blockers was not associated with a reduction in all-cause mortality in patients with MI without heart failure or systolic dysfunction. This finding suggests that not all patients with MI should benefit from ß-blocker treatment, particularly those without heart failure or systolic dysfunction. Randomised controlled trials are needed to confirm this result [173].

On the basis of current evidence, the routine administration of beta-blockers in all post-STEMI patients should be considered. Beta-blockers are recommended in patients with reduced left ventricular systolic function (LVEF <_40%), in the absence of contraindications such as acute heart failure, haemodynamic instability or high-grade AVB [129].

As no study has adequately addressed the duration of beta-blocker therapy to date, no recommendation has been made in this regard. With regard to the timing of initiation of oral beta-blocker therapy in patients who do not receive an early IV beta-blocker, a retrospective analysis of 5259 patients suggested that early (i.e. <24 h) administration of a beta-blocker provides a survival benefit over delayed administration [174]. Therefore, in haemodynamically stable patients, initiation of an oral beta-blocker should be considered within the first 24 hours.

f. Converting enzyme inhibitors

A systematic review of the use of ACE inhibitors in early STEMI indicates that this treatment is safe, well tolerated and associated with a small but significant reduction in 30-day mortality, with most of the benefit being seen in the first week after the acute event [175]. The use of ACE inhibitors after MI has been associated with a significant reduction in mortality and incidence of reinfarction [176]. However, data on ACE inhibitors in patients treated with primary PCI are scarce. The mechanisms by which ACE inhibitors reduce adverse events after MI include reducing ventricular remodelling after STEMI, decreasing sympathetic activity and increasing vagal tone, which may reduce the incidence of sudden death [177].

Treatment with ACE inhibitors is recommended in patients with left ventricular systolic dysfunction, heart failure, hypertension or diabetes, and should be considered in all STEMI patients [178, 179]. Patients who cannot tolerate a converting enzyme inhibitor should receive an angiotensin II receptor blocker (ARB). In STEMI, valsartan was shown to be non-inferior to captopril in the VALIANT (VALsartan In Acute myocardial iNfarcTion) trial [180].

g. Lipid-lowering treatment

Lipid-lowering treatment is one of the most important elements of secondary prevention. Statins, which reduce circulating cholesterol levels, are a well-established treatment for the secondary prevention of cardiovascular disease, as demonstrated by several randomised controlled trials [181-183]. A meta-analysis of over 90,000 people in 14 randomised trials confirms the benefit of statin use after hospital admission for MI, with a reduction in total mortality as well as other coronary events [184].

Further evidence is provided by a study of more than 105,000 patients who began treatment with a moderate or high intensity statin after hospitalisation for myocardial infarction; overall, 52.8% of patients had high adherence to statin treatment and only 1.7% had statin intolerance. The results suggest a 36% higher incidence of recurrent MI in statin intolerant patients and a 43% higher risk of CHD in statin intolerant patients compared to those with high statin adherence [185], while the risk of mortality was similar in both groups over a median follow-up period of 1.9-2.3 years. Lipid-lowering treatment should be started as early as possible, as this increases patient compliance after hospital discharge, and should be based on high-intensity satins, because of the early and lasting clinical benefits [129].

The use of lower intensity statin therapy should be considered in patients at increased risk of statin side effects (e.g. elderly, hepatic or renal impairment, previous side effects or potential interaction with essential concomitant therapy). The goal of treatment is an LDL-C concentration of <0.55 mmol/L or at least a 50% reduction in LDL-C [186, 187]. In patients who cannot tolerate statins, treatment with ezetimibe should be considered.

Genetic studies have shown lower levels of LDL cholesterol and a reduced risk of CVD in people with loss-of-function mutations in PCSK9, the gene that codes for proprotein convertase subtilisin/kexin type 9, a protein that promotes catabolism of LDL receptors, thereby increasing circulating levels of LDL particles [188].

Recommended therapeutic targets for LDL cholesterol	
Risk category	**LDL-C target values (start with untreated LDL-C)**
Very high	**<1.4 mmol/1 / 0.55 g/1 and decrease >50%.**
High	**<1.8 mmol/1 / 0.70 g/1 and >50% decrease**
Moderate	**<2.6 mmol/1/ 1.00 g/1**
Low	**<3.0 mmol/1 / 1.16 g/1**

Figure 14: Recommended therapeutic targets for LDL-C

Drugs targeting PCSK9 include monoclonal antibodies (evolocumab and alirocumab) and small interfering RNAs (inclisiran) and are associated with impressive reductions (40-60%) in LDL cholesterol when combined with statin therapy [189, 190]. A randomised controlled trial showed a 20% reduction in the relative risk of cardiovascular events (cardiovascular death, MI or stroke) after a median follow-up of 2.2 years [189], and a trial of alirocumab showed a 15% reduction in the relative risk of major cardiovascular events [191]. However, these drugs are not currently used routinely in STEMI, largely because of their high cost.

TableIII: Intensity of lipid-lowering therapies [192].

Treatments	Discount theoretical LDL-c
Statin moderate **-ROSUVASTATIN 5mg** **-ATORVASTATINE 10mg** **-SIMVASTATINE 20mg-40mg** **-FLUVASTATIN 80mg**	**30%**
Statin forte **-ROSUVASTATIN 10mg-20mg-40mg** **-ATORVASTATIN 20mg-40mg-80mg**	**50%**
STATINE Forte + ezetimibe	**65%**
PCSK9 inhibitor	**60%**
Strong statin + a PCSK9 inhibitor	**75%**
Strong statin + an inhibitor + ezetimibe	**85%**

h. Calcium inhibitors

A meta-analysis of 17 trials of CIs in the acute phase of STEMI showed no beneficial effect on mortality or reinfarction, with a trend towards higher mortality in patients treated with nifedipine. Therefore, the routine use of CIs in the acute phase is not indicated [193, 194].

In the chronic phase, a randomised controlled trial of 1,775 patients with myocardial infarction not on beta-blockers who were treated with verapamil or placebo showed that the risk of mortality and reinfarction was reduced with verapamil [195]. Thus, in patients with contraindications to beta-blockers, particularly in the presence of obstructive airway disease, CIs are an option for patients without heart failure or impaired LV function.

Systematic use of dihydropyridines, on the other hand, has not shown any benefit after STEMI [196], and they should therefore only be prescribed if there are clear additional indications, such as hypertension or residual angina [197].

i. Nitro derivatives :

The routine use of nitrates in STEMI did not provide any benefit in a randomised placebo-controlled trial and is therefore not recommended [198]. Intravenous nitrates may be useful during the acute phase in patients with hypertension or heart failure, provided there is no hypotension, VD infarction or use of PDE5 inhibitors in the preceding 48 hours.

After the acute phase, nitrate derivatives remain valuable agents for controlling the residual symptoms of angina pectoris.

j. Mineralocorticoid receptor antagonists / Anti-aldosterone :

Treatment with mineralocorticoid receptor antagonists is recommended in patients with left ventricular dysfunction (LVEF <_40%) and heart failure after STEMI [199, 200].

Eplerenone, a selective aldosterone receptor antagonist, has been shown to reduce morbidity and mortality in these patients. The Eplerenone PostAMI Heart failure Efficacy and SUrvival Study (EPHESUS) involved 6642 post-MI patients with left ventricular dysfunction (LVEF <_40%) and symptoms of heart failure/diabetes who were started on eplerenone within 3 to 14 days of MI [199]. After a mean follow-up of 16 months, there was a 15% relative reduction in total mortality and a 13% reduction in death and hospitalisation for cardiovascular events.

Two studies have also shown a beneficial effect of early treatment with MRA in STEMI without heart failure. The REMINDER trial (Evaluating The Safety And Efficacy Of Early Treatment With Eplerenone In Patients With Acute Myocardial Infarction) involved 1012 STEMI patients without heart failure who were treated with eplerenone or placebo within 24 hours of the onset of symptoms [201]. The ALBATROSS trial (Aldosterone Lethal effects Blockade in Acute myocardial infarction Treated with or without Reperfusion to improve Outcome and Survival at Six months) randomised 1603 patients to an IV bolus of potassium canrenate (200 mg) followed by spironolactone (25 mg per day) vs. placebo [202].

Future studies will clarify the role of MRA treatment in this context.

k. New strategies targeting inflammation :

Inflammation plays an important role in atherogenesis and plaque progression [203]. The CANTOS trial was the first to validate the inflammatory hypothesis in a large cohort study of patients with coronary artery disease: targeting the IL-1ß immune pathway with canakinumab (a human anti-IL-1ß monoclonal antibody) resulted in a clinically significant 15% relative reduction in major cardiovascular events compared with placebo, independent of LDL levels [204]. Anakinra, an IL-1 receptor inhibitor, has been evaluated in two small phase II studies in patients with acute MI and has been shown to reduce CRP levels [205, 206].

In addition to the interleukin pathway, T-cell activation signalling, synthetic protein tyrosine phosphatase inhibitors and low doses of IL-2 are under development or represent future areas of research to target inflammation in ACS [207].

l. Management of "No-reflow" and reperfusion lesions :

Certain aspects of the pathophysiology of infarction remain surprisingly refractory even after well-managed and successful treatment. No-reflow" is a complication that can occur after coronary revascularisation and is associated with an increased risk of cardiogenic shock, recurrent infarction and mortality [208, 209].

Although the risk factors are well known (such as high thrombus burden, the context of primary PCI per se and ischaemic time), there are still no truly effective treatment options. The use of GPIIb/IIIa inhibitors and vasodilators has failed to demonstrate clinical benefit in several observational studies and randomised trials [210]. Reducing the total ischaemic time by getting the patient to the cardiac catheterisation room as quickly as possible remains one of the most effective ways of preventing the "No-reflow" phenomenon and reperfusion lesions. To this end, devices that alert the patient to the presence of an infarct can reduce the time between the onset of symptoms and admission to hospital and, consequently, total ischaemic time [211, 212].

Although drug trials targeting reperfusion injury have not shown beneficial results in terms of clinical outcomes, there is evidence that remote ischaemic conditioning has a protective effect on the creur and improves clinical outcomes by reducing myocardial injury [213]. A better understanding of the complex pathophysiology, including the mechanism of reperfusion injury, may identify additional targets to prevent loss of microvascular integrity and increase vascular permeability. Despite the many failures to date, the prevention and treatment of reperfusion lesions, which directly affect infarct size, should remain the focus of future cardiovascular research.

m. Therapies to reduce infarct size and microvascular obstruction:

Final infarct size and microvascular obstruction are major independent predictors of long-term mortality and heart failure in STEMI survivors [214, 215]. MVO is defined as inadequate myocardial perfusion after successful mechanical opening of the ARF, and is caused by several factors [216]. MVO is diagnosed immediately after PCI when the post-procedure angiographic TIMI flow is <3, or in the case of a TIMI flow of 3 when the degree of myocardial reddening is 0 or 1, or when ST resolution within 60-90 minutes of the procedure is <70%.

Other non-invasive techniques for diagnosing MVO include late gadolinium enhancement in cardiac MRI (the current state-of-the-art technique for identifying and quantifying MVO), contrast and magnetic resonance ultrasound, single-photon emission computed tomography (SPECT) and positron emission tomography (PET) [216]. Different strategies, such as coronary post-conditioning, remote ischaemic conditioning, early i.v. metoprolol, GP inhibitors, etc., have been proposed. GP IIb/IIIa inhibitors, drugs targeting mitochondrial integrity or nitric oxide pathways, adenosine, glucose modulators, hypothermia and others, have been shown to be beneficial in preclinical and small-scale clinical trials [215, 217], but there is still no therapy aimed at reducing ischaemia/reperfusion injury (MI size) that is clearly associated with improved clinical outcomes. Reducing ischaemia/reperfusion injury in general, and microvascular obstruction in particular, remains an unmet need to improve long-term ventricular function in STEMI.

9. Complications of STEMI :

a. Supraventricular arrhythmias :

The most common supraventricular arrhythmia is atrial fibrillation (AF), affecting 21% of STEMI patients [218]. AF may be pre-existing, detected for the first time or of recent onset. Patients with AF have more co-morbidities and a higher risk of complications [219].

In many cases, the arrhythmia is well tolerated and no specific treatment is required, apart from anticoagulation [220]. Rapid treatment is necessary in cases of acute haemodynamic instabilityê. There is little evidence to suggest a preference for rate control over rhythm control in this situation [221]. Electrical cardioversion should be considered but early recurrence of AF is common after successful cardioversion.

Control of acute rhythm with antiarrhythmic drugs is limited to the use of amiodarone [218, 220]. Adequate control of heart rate can be achieved with beta-blockers [221, 222].

Several studies, but not all, have suggested that new-onset AF can be reduced by beta-blockers, ACE inhibitors/ARB2, and also by early-onset statin therapy [218].

Patients with AF and risk factors for thromboembolism should be treated appropriately with chronic oral anticoagulation [220]. STEMI patients with documented AF have a worse short- and long-term prognosis than patients in sinus rhythm [219, 223]. The presence of AF is associated with a higher rate of re-infarction, a higher rate of stroke, a higher risk of heart failure, and may also increase the risk of sudden cardiac death [218, 219, 224]. It should be noted that AF during STEMI is associated with a significantly higher rate of stroke at long-term follow-up [219, 224].

b. Ventricular arrhythmias :

The incidence of ventricular tachycardia (VT) and ventricular fibrillation (VF) after

STEMI has decreased since the establishment of routine early reperfusion and treatment with ß-blockers [225]. However, recent studies report an incidence rate of nearly 6% of VT or VF in patients with MI [226, 227].

Urgent reperfusion is very important because ischaemia often triggers these arrhythmias [228]. Beta-blockers are recommended in the absence of contraindications [229, 230]. Repetitive electrical cardioversion or defibrillation may be necessary [231]. If there is insufficient control, I.V. administration of amiodarone is recommended [227, 232]. If amiodarone is contraindicated, I.V. administration of lidocaine may be considered.

Ventricular tachycardia and ventricular fibrillation occurring early in the 48 hours after STEMI appear to be associated with increased in-hospital mortality but do not appear to affect long-term prognosis [233, 234]. Conversely, VT or VF that develops after 48 hours and in the absence of recurrent ischaemia is associated with a worse prognosis, requiring aggressive treatment as well as evaluation of the use of an implantable cardioverter defibrillator (ICD) [235].

c. Sinus bradycardia and atrioventricular block :

As with ventricular tachyarrhythmias, the occurrence of conduction disturbances and bradyarrhythmias associated with acute myocardial infarction has decreased in the era of early revascularisation [236]. More specifically, the incidence of atrioventricular block in STEMI patients treated with thrombolysis is 6.9% [237], compared with an incidence rate of 3.2% in patients treated with primary PCI [238].

Sinus bradycardia is mainly associated with STEMI of the lower wall of the heart and can occur in the first hours and up to several days after the infarction. Sinus bradycardia and various degrees of atrioventricular block occurring in the early phase of MI due to increased vagal tone respond well to atropine and usually correct within 24 hours [239]. Subsequent development of conduction delay may be associated with redema and local accumulation of adenosine [240].

These abnormalities are generally asymptomatic but can also lead to haemodynamic instability. In the case of haemodynamic instability, temporary pacing (Pacemaker) may be required; however, most conduction abnormalities due to inferior territory STEMI resolve within 2 weeks [236].

Atrioventricular block associated with an anterior infarct is most often infra-hisian (located lower down the conduction pathway), with a new bundle branch block or hemiblock usually indicating an extensive anterior infarct [129, 236].

In-hospital mortality and 30-day mortality are higher in patients with VAD than in those without VAD, irrespective of the location of the infarction, while outcomes beyond 30 days appear similar between the groups [241].

d. Left ventricular thrombus :

Data on the incidence of left ventricular thrombus detected by optimal imaging methods range from 15% in STEMI patients and up to 25% in patients with previous MI [242].

Consistent with previous studies, the risk factors for left ventricular thrombus formation are prior STEMI and LVEF less than 50% [243, 244]. The presence of microvascular obstruction [243] and apical wall motion abnormality [245] has also been associated with the development of left ventricular thrombus.

Despite the low sensitivity of standard transthoracic echocardiography, it remains more accessible and affordable than cardiac MRI. The apical wall motion score has been shown to be useful in identifying individuals who would benefit from cardiac MRI for the detection of left ventricular thrombus. As a result, an algorithm [246] has been proposed for application at the time of primary PCI, which can be used to identify those most likely to develop left ventricular thrombus and who may benefit from cardiac MRI if TTE is inconclusive.

Gellen et al [247] recently showed that the highest rate of detection of left ventricular thrombus (25%) was in those who underwent cardiac MRI between 9 and 12 days after a previous STEMI, Meurin et al [248] made similar findings, with a higher incidence of left ventricular thrombus in patients who underwent cardiac MRI between 8 and 15 days. These data suggest that the optimal time for imaging detection of left ventricular thrombus after acute STEMI may be at 2 weeks.

Once a left ventricular thrombus has been detected, anticoagulant treatment (in addition to dual anti-platelet therapy) is essential, although it increases the risk of haemorrhage.

The large proportion of left ventricular thrombi (88%) resolved between 3 and 6 months. Follow-up with repeated cardiac MRI from 3 months after STEMI may reduce the need for prolonged anticoagulation in a proportion of patients.

Zhang Z et al [249] in a recent study on the prevention of left ventricular thrombus concluded that the addition of low-dose rivaroxaban (2.5 mg twice daily for 30 days) to dual antiplatelet therapy could prevent left ventricular thrombus formation within 30 days in patients with previous STEMI after primary percutaneous coronary intervention. However, larger trials are needed to investigate these findings further.

e. Mechanical complications.

Mechanical complications after acute infarction include papillary muscle rupture and dysfunction, left ventricular aneurysm, interventricular septal rupture and LV free wall rupture. These complications have become rare events following the advent of primary PCI, and their current incidence is estimated at

<1%. A report from the APEX-AMI study showed rates of 0.51% for rupture of the free wall, 0.17% for rupture of the interventricular septum and 0.26% for rupture of the papillary muscle [250]. Nevertheless, these complications constitute a life-threatening emergency scenario, requiring early diagnosis and urgent surgical intervention. Initial symptoms range from dyspnoea to fulminant heart failure, cardiogenic shock and sudden cardiac death. Each of these symptoms, as well as the appearance of a new heart murmur, should always prompt an immediate echocardiogram, which is the gold standard for diagnosing mechanical complications. In addition, transfer to the intensive care unit is mandatory.

Once the diagnosis has been confirmed, subsequent treatment and surgical planning should be assessed by a multidisciplinary team. In cases of haemodynamic instability or cardiogenic shock, the insertion of an intra-aortic counterpulsation balloon may be considered, while refractory circulatory failure should prompt the introduction of extracorporeal membrane oxygenation (ECMO) as rescue therapy to manage the patient until definitive treatment is instituted [251, 252].

f. Cardiogenic shock :

Cardiogenic shock occurs in 5-15% of patients, is one of the strongest predictors of short- and long-term prognosis after STEMI [253-255] and remains the leading cause of death after acute MI [255]. Mechanical complications are rare causes of cardiogenic shock, as the most common underlying cause is left ventricular dysfunction. The results of randomised controlled trials in patients with cardiogenic shock are limited. Long-term follow-up from the SHOCK trial showed that early reperfusion was associated with greater survival compared with initial medical stabilisation [256].

10. Prevention of atheromatous coronary artery disease :

According to WHO recommendations, there are several stages of prevention: **primary prevention, secondary prevention and tertiary prevention.**

The prevention of atheromatous coronary artery disease is mainly based on the management of the DRFs and includes three methods at each stage:

- **Information**, through major public health campaigns;
- **Training in public health**;
- **Health education**.

Cardiovascular prevention can be **individual, by** assessing all the risk factors, or **collective, by anti-smoking regulations, limiting salt and sugar content in the** food **industry**, school education and nutrition, informing the population through various national or regional campaigns, and providing access to sports facilities and outdoor areas... [257, 258].

The fight against smoking and hygienic and dietetic measures can provide better results than drug treatments (lipid-lowering drugs).

a. Primary prevention :

The aim of primary prevention is to reduce **the incidence of atheromatous cardiovascular disease**. Every consultation should be an opportunity to raise the subject of prevention and to gather information on the main risk factors, diet and physical activity. The same applies to a baseline biological check-up (including fasting blood glucose and lipid levels) to detect metabolic abnormalities.

On the basis of these data, the SVR should be assessed and advice and treatment adapted to the level of risk.

Primary prevention of CVD remains **focused on changing individual risk behaviours.** Management of CVD includes smoking cessation, dietary measures, advice on physical activity and appropriate treatment for each FDR (hypertension, hyperlipaemia, diabetes, etc.).

b. Secondary prevention :

Secondary prevention is the set of actions designed to **reduce the prevalence of CVD**, and therefore the time it takes to develop. It is involved **in screening** and includes **treatment of** the disease. It must **be systematic, multi-risk** (all risk factors must be taken into account), **intensive** (optimal for each factor) and **medicalised** (the use of medication is the rule, with regular medical follow-up).

The prescription of treatment must be accompanied by insistent advice on lifestyle changes. The fight against risk factors becomes even **more imperative** than in primary prevention, and the objectives of primary prevention **are stricter**.

c. Tertiary prevention :

This is the set of measures designed to prevent the onset of complications, after-effects or handicaps linked to the disease. It is based on :

- treatment of complications (thromboembolism, rhythm disorders, etc.).

- specialised cardiology rehabilitation.
- palliative care: social reintegration and psychological care.

II. FIBRINOLYTICS Fibrinolytics

A. Fibrinolysis :

1. Definition :

Fibrinolysis is defined as the process of degradation of the fibrin network, an essential component of thrombus, by plasmin. When a clot forms, a physiological fibrinolysis mechanism is set in motion to destroy the fibrin network and restore vascular flow as quickly and completely as possible. It is based on hydrolysis of the peptide bond which links the two chames of plasminogen, an inactive proenzyme of hepatic origin, to plasmin, unmasking the catalytic site of this powerful proteolytic enzyme. In fact, plasminogen has a strong affinity for fibrin polymers, which gives it the ability to dissolve fibrin preferentially when converted to plasmin. However, this enzyme is not selective and can degrade both fibrin and fibrinogen, as well as certain coagulation factors (V, VIII).

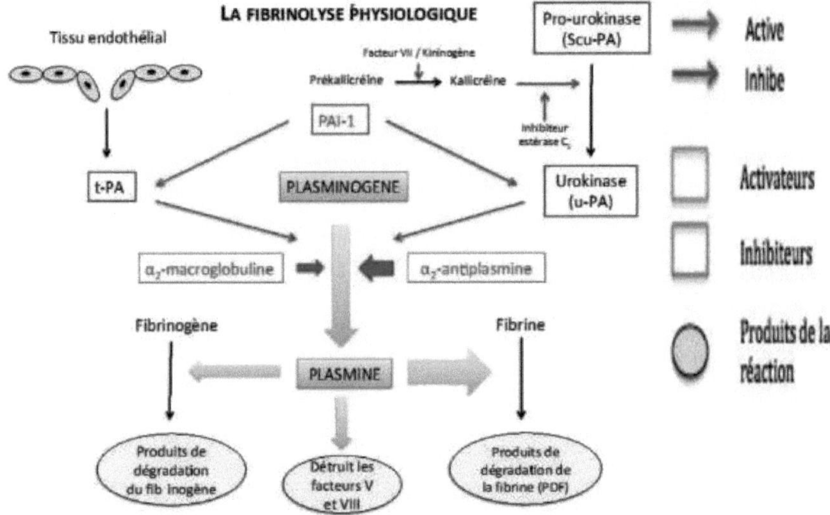

Figure 15: Diagrams of physiological fibrinolysis

This physiological process involves modulators (Figure 15):

- Plasminogen activators such as t-PA (synthesised by the cells of the

and released during thrombus formation) and u-PA (urokinase, produced by the kidney from scu-PA, or pro-urokinase, and excreted in the urine);

- Inhibitors such as plasma inhibitors of plasmin (a2-

antiplasmin, rapid antiplasmin, and a2-macroglobulin) as well as PAI-1 inhibiting plasminogen activators (t-PA, u-PA).

In therapy, there are organic inhibitors of fibrinolysis, such as aprotinin extracted from sheep lung or pancreas, as well as chemical inhibitors such as epsilon-aminocaproic acid and tranexamic acid.

2. Studies on fibrinolysis in STEMI :

a. The benefits of early reperfusion :

It is now established that thrombolysis reduces mortality and improves the prognosis of patients treated before the twelfth hour of infarction. The earlier the treatment is started, the greater the beneficial effect. The GUSTO studies demonstrated that normalisation of coronary flow is a major criterion of survival and that the prognosis of thrombolysed myocardial infarction depends on the time factor in the speed of coronary recanalisation and downstream myocardial tissue reperfusion. The GUSTO study clearly shows that a one-hour reduction in the time taken to initiate thrombolysis saves five extra lives per thousand patients treated.

The BOERSMA meta-analysis included twenty-two randomised trials from 1983 to 1993 (50,246 patients) and showed that the reduction in mortality at thirty-five days was -44% in patients thrombolysed within two hours compared with -20% in those thrombolysed later; it also reported sixty-five, thirty-seven and twenty-six lives saved per thousand patients treated at the first, second and third hour respectively [259]. In the GISSI 1 study, the reduction in mortality was 50% in patients thrombolyzed before the first hour.

As a logical consequence of these data, it soon became apparent that developing pre-hospital thrombolysis was one of the most effective ways of reducing the time taken to administer thrombolytics. Two large randomised trials, one European (EMIP) and the other American (MITI), compared pre-hospital thrombolysis with hospital thrombolysis. They demonstrated a significant reduction in mortality of 15% and a time saving of around fifteen minutes [260, 261]. In a study of 6,607 patients, BOERSMA demonstrated a time saving of one hour for pre-hospital thrombolysis compared with hospital fibrinolysis.

Even if it is true that the maximum benefit of thrombolysis is obtained in the very first hours of a myocardial infarction, we suspected that a benefit could be envisaged beyond the sixth hour of the myocardial infarction. This was already suggested in ISIS-2, since patients were included up to the twenty-fourth hour and a benefit was observed in the sub-group of patients treated between six and twenty-four hours. Later, in the LATE study [262], a significant reduction in mortality was observed in patients treated with thrombolysis between six and twelve hours, but not beyond.

b. Open artery theory :

These data led to the development of the "open artery theory", meaning that coronary flow had to be restored rapidly and completely. The GUSTO study validated this theory by establishing the link between the degree of coronary patency and improved prognosis. This study also showed that the earlier, more complete and longer-lasting the repermeabilisation, the better the preservation of ventricular function and the lower the mortality [263].

c. Mortality as the main criterion for judging thrombolytics :

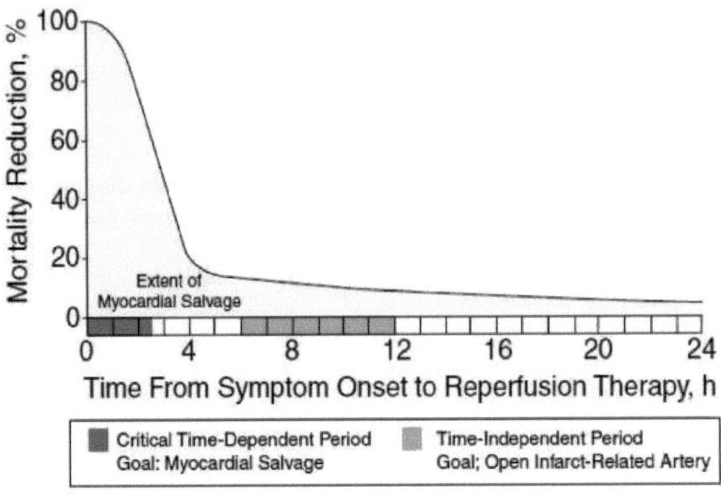

Figure 16: Reduction in mortality as a function of total ischaemic time

It should be noted that the extent of myocardial recovery decreases considerably in the first 4 hours, but that up to 24 hours, recovery of a smaller proportion of myocardium is still possible.

Major studies have made it possible to define the advantages and disadvantages of different thrombolytic substances and to validate the best strategies or therapeutic regimens in terms of efficacy and safety.

GISSI-1 compared conventional treatment with intravenous streptokinase in patients with myocardial infarction of less than twelve hours, and showed a reduction in hospital mortality from 13% to 10.7% with thrombolytics.

> ISIS-2 demonstrated the potentiation of the beneficial effect of streptokinase by the addition of aspirin and a 39% reduction in hospital mortality.

> AIMS showed a 47% reduction in mortality at one month on APSAC compared with placebo in 1258 patients admitted for an infarction of less than six hours.

> ASSET included 5011 patients aged under 75 randomised within the first five hours of an infarction and showed a 26% reduction in mortality at one month on t-PA compared with placebo.

> GISSI-2 and ISIS-3 are two major studies comparing first and second generation thrombolytics. The first study compared t-PA and streptokinase in infarction of less than six hours and showed no significant difference between the two groups in terms of 35-day mortality and major cardiovascular events. The second study compared streptokinase, t-PA (dutéplase) and APSAC in infarction of less than twenty-four hours and showed that there was no significant difference in mortality between the three treatment groups **[264, 265]**.

> Thanks to their concept, the GUSTO studies initiated in the nineties have provided decisive answers, with a high level of scientific evidence, to a number of unanswered

questions (superiority of certain thrombolytics over others, choice of antithrombin to combine, therapeutic regimes or protocols, etc.). The following are described:

S GUSTO-I: this study validated the open artery hypothesis and definitively established the superiority of accelerated t-PA combined with intravenous heparin over the three other protocols (streptokinase and subcutaneous heparin, streptokinase and intravenous heparin and finally a combination of the two thrombolytics with a lower dose of streptokinase) by demonstrating a significant reduction in thirty-day mortality (6.3% vs. 7.3%). It was not until this study that it was demonstrated that the prognosis of myocardial infarction depended essentially on the preservation of left ventricular function, which in turn depended on the rapidity, quality and durability of repermeation [266].

S GUSTO 2 B: demonstrated the ineffectiveness of hirudin in terms of reducing the occurrence of major cardiac events, but with a higher haemorrhagic risk [267].

S GUSTO 3: compared alteplase (second generation) in an accelerated protocol with reteplase (third generation) in a double intravenous bolus. There was no significant difference in terms of mortality or incidence of stroke [268].

d. Other assessment criteria :

Other thrombolysis assessment criteria are also taken into account.

> The patency of the artery presumed to be responsible for the myocardial infarction (PRA): this illustrates the thrombolytic's ability to lyse the occlusive clot. It is judged on the basis of a "qualitative" TIMI grade analysis of coronary flow during coronary angiography performed at 90 minutes. Only TIMI grade 3 (complete opacification without delay of the RPA) is considered pathognomonic of recanalisation. The rates of complete recanalisation (TIMI 3) are around 30% with streptokinase, 54% with "accelerated" t-PA and 60% with double-bolus reteplase.

More recently, the TIMI 10A, TIMI 10B, ASSENT-1 and ASSENT-2 studies using a third-generation thrombolytic, TNK-tPA, have demonstrated, in addition to the ease of single-bolus administration of this substance, a similarity in TIMI 3 reperfusion rates compared with t-PA, an improved tolerance profile and equivalence between TNK-tPA and t-PA in terms of 30-day mortality [269-272].

> Left ventricular function (LVF): GUSTO angiography has shown that there is a strong correlation between coronary patency and preservation of left ventricular function. Furthermore, long-term maintenance of coronary patency limits post-myocardial infarction ventricular remodelling.

> The concept of net clinical benefit: emerged with the GUSTO studies. It is the combination of mortality at D30 and disabling stroke, including cerebral haemorrhage. It defines a group of patients for whom thrombolysis was not life-saving or resulted in a disabling or fatal stroke. This concept of net clinical benefit soon appeared to be more appropriate and judicious than that of total mortality.

G. Recommendations :

If primary PCI is not available within 120 minutes of diagnosis, fibrinolysis should be performed, unless contraindicated.

The very large US National Registry of Myocardial Infarction (NRMI) reported that the mortality advantage of primary PCI over on-site fibrinolysis disappeared when the time from first medical contact to PCI exceeded 120 minutes [273].

The guidelines recommend the use of fibrinolysis for STEMI patients who present within

12 hours of symptom onset and in whom PCI cannot be performed within 120 minutes. Treatment should then be started within 30 minutes of the first medical contact, after careful assessment of any contraindications.

Guidelines generally favour fibrin-specific fibrinolytic agents (tenecteplase, alteplase or reteplase), with a specific dosage according to age and weight [129, 140].

It is important to note that even if fibrinolysis is performed as a primary therapeutic approach, it must always be followed by coronary angiography and possibly PCI. The time taken to perform coronary angiography depends on the success of fibrinolysis, based on certain criteria.

When the diagnosis of STEMI+ is made less than 3 hours after the onset of symptoms, primary PCI is the strategy of choice. If the anticipated time from diagnosis of ST+-MI to PCI-mediated reperfusion is >120 minutes, then immediate fibrinolysis is indicated. After 3 hours (and up to 12 hours) from the onset of symptoms, the later the patient presents, the more preference should be given to the primary PCI strategy over fibrinolysis.

In the case of advanced ST+ MI (12 to 48 hours after the onset of symptoms), a strategy of routine primary PCI (coronary angiography and PCI if indicated) should be considered in all patients. After 48 hours (recent ST+ MI), coronary angiography should be performed, but routine PCI of a fully occluded MI artery is not recommended. Regardless of the length of time since onset of symptoms, the presence of symptoms suggestive of ischaemia, haemodynamic instability or life-threatening arrhythmias is an indication for a primary PCI strategy.

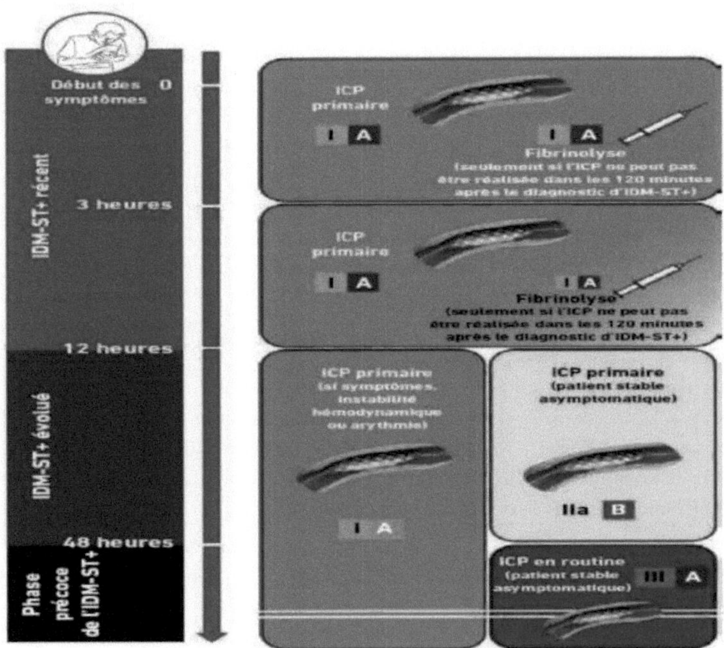

Figure 17: Strategies for reperfusion of the infarct artery according to time since onset of symptoms

The later the patient presents (particularly after 3 hours) [273-275], the more consideration should be given to transfer for primary PCI (rather than fibrinolytic therapy) as the efficacy and clinical benefit of fibrinolysis decreases as the time to onset of symptoms increases [273].

- The aim of fibrinolytic treatment is to dissolve the thrombus by activating plasminogen, which leads to the formation of plasmin, which cleaves the fibrin bonds within the thrombus. Alteplase, tenecteplase and reteplase are fibrin-specific agents, which means that they preferentially bind to fibrin in a thrombus and catalyse the cleavage of trapped plasminogen into plasmin [276], resulting in localised fibrinolysis with limited systemic proteolysis. Conversely, streptokinase is a non-specific agent for fibrin and causes an indirect change in the plasminogen molecule, which then acts as a plasmin [276].

B. Mode of action :

1. Pharmacological **properties** :

a. **Molecules** from 1re generation:

Streptokinase is a highly purified protein obtained from culture filtrates of group C β-haemolytic streptococci. By combining with plasminogen, streptokinase (like urokinase) forms an activator complex that metabolises plasminogen to plasmin, enabling fibrinolysis and fibrinogenolysis.

b. **2nd** generation **molecules** :

Alteplase is a glycoprotein responsible for degrading plasminogen to plasmin by selectively binding to fibrin, thus enabling selective fibrinolysis.

c. **3rd generation molecules** :

Reteplase and tenecteplase are recombinant proteins derived from t-PA that selectively catalyse, in the presence of fibrin, the cleavage reaction of thrombus-bound plasminogen into plasmin, which degrades the fibrin matrix of the thrombus. Compared with endogenous t-PA, tenecteplase is more specific for fibrin and more resistant to inactivation by its endogenous inhibitor (PAI-1).

2. **Pharmacokinetics** :

Fibrinolytics are administered intravenously (or intra-arterially). The half-life (table 1) is a parameter that must be taken into account insofar as it reflects the duration of the benefits, but also of the risks associated with the use of the product, and therefore plays a role in the choice of a compound for a specific indication. Streptokinase, for example, has a long half-life, which increases the risk of haemorrhage, while urokinase has an extremely short half-life, requiring infusion.

Table IV: Different fibrinolytics and protocols recommended by the European Society of Cardiology (ESC 2017)

Fibrinolytic	Dose
Streptokinase	**1.5 million units in 30-60 min IV (allergy risk)**
Alteplase (tPA)	**15 mg IV bolus 0.75 mg/kg IV over 30 min (maximum 50 mg) Then 0.5 mg/kg IV over 60 min (maximum 35 mg)**
Reteplase (tPA)	**10 units IV bolus + 10 units 30 min later**
Tenecteplase (TNK-tPA)	**A single IV bolus : 30 mg (6000UI) if < 60 kg 35 mg (7000IU) if < 70 kg 40 mg (8000IU) if < 80 kg**

	45 mg (9000IU) if < 90 kg 50 mg (10000IU) if > 90 kg It is recommended that the dose be reduced by half in patients aged 75 and over.

3. **Presentations :**

- **Tenecteplase (Metalysis)** : Box containing: 1 x 20 ml powder vial with flip-off cap + 1 x 10 ml pre-filled syringe with sterile adapter.

Each vial contains 10,000 units (50 mg) of tenecteplase.

1ml of reconstituted solution contains 1,000 units (5 mg) of Tenectéplase.

Tenectéplase is a fibrino-specific plasminogen activator produced by the recombinant DNA technique in a Chinese hamster ovary cell line.

As a rule, Tenecteplase should be used for patients admitted before the 6th hour. A dose adapted to weight is diluted in physiological saline and administered over 10 seconds.

- **t-PA (Alteplase):** dosed at 50 MG powder + solvent (50 ML). It is administered according to GUSTO protocol III (accelerated protocol), i.e. an IV bolus of 15 MG then 50 MG to be administered over thirty minutes then 35 MG to be administered over one hour. The total dose administered must not exceed 100 MG regardless of weight.

C. Indications:

- **Endovascular treatment** (streptokinase).
- **Desobstruction of arteriovenous shunts** in haemodialysis patients with indwelling catheters (streptokinase).
- **Pulmonary embolism** (urokinase, alteplase).
- Recent **arterial or venous occlusions** (urokinase).
- Acute **myocardial infarction** (< 6 hours: tenecteplase; < 12 hours: alteplase, reteplase).
- Acute stroke (< 4H30 hours: alteplase).

D. Contraindications to fibrinolysis (ESC 2017):

♦ **Contraindications for fibrinolytics focus on preventing the risk of bleeding:**

In the presence of contraindications to fibrinolytic therapy, it is important to weigh the potentially life-saving effect of fibrinolysis against the potentially fatal side-effects, taking into account alternative treatment options such as **delayed primary PCI.**

Table V: Absolute contraindications to thrombolysis

History of intracranial haemorrhage or stroke of unknown origin
Ischaemic stroke within the last 6 months
Central nervous system lesion, tumour or arteriovenous malformation
Major trauma / surgery / head injury in the previous month
Gastrointestinal haemorrhage in the previous month
Known haemostasis disorder
Dissection of the aorta
Non-compressible punctures in the last 24 hours (liver biopsy, lumbar puncture, etc.)

TableVI: Relative contraindications to thrombolysis

Relative contraindications to thrombolysis

TIA within the last 6 months
Oral anticoagulant treatment
Active pregnancy or 1st week post partum

Refractory hypertension (SAP > 180 mmHg and/or DBP > 110 mmHg)
Advanced liver disease
Infectious endocarditis
Active peptic ulcer

Prolonged or traumatic resuscitation

+ Precautions for use relating to fibrinolytics must be observed:

- The use of rigid catheters or recent intra-muscular injections (within 48 hours) should be avoided;
- Particular care should be taken when administering fibrinolytics to elderly patients (> 75 years) and to low-weight hypertensive women;
- Anti-arrhythmic treatment must be available at the time of administration.

as well as resuscitation equipment;

- It is recommended that fibrinolytics should not be used in patients under 18 years of age, due to a lack of studies on safety and efficacy (except for alteplase, for which use is limited);
- Fibrinogen and thrombin levels must be monitored.
- **/Special cases: Pregnancy and breastfeeding:**

4 **The administration of fibrinolytics to pregnant women** represents a major haemorrhagic risk, particularly when injected at the time of delivery. In addition, animal studies have shown that urokinase has a teratogenic effect, and urokinase is therefore not recommended for thrombo-embolic events in pregnant women. Conversely, reteplase and streptokinase do not cross the placental barrier at the end of pregnancy or during labour.

♦**Fibrinolytics have not been shown to be excreted in breast milk,** but breast milk should not be used for 24 hours after thrombolysis as a precaution.

E. Undesirable effects :

As all fibrin deposits are lysed by fibrinolytics, whether they are part of a pathological or haemostatic thrombus, the major iatrogenic risk is mainly haemorrhagic, particularly in the brain.

♦ **Haemorrhagic disorders**: haemorrhage at the site of injection, haematoma, epistaxis, haematuria, ecchymosis, intracranial haemorrhage (particularly when treating acute stroke), haemorrhage from the digestive, respiratory and urogenital tracts.

♦ **Cardiac disorders:** recurrent myocardial ischaemia, heart failure, reperfusion arrhythmia.

♦ **Vascular disorders**: embolism, hypotension.

♦ **Rare allergic reactions:** urticaria, erythema, arthralgia, arthritis, convulsions, bronchospasm, redema, hypotension, myalgia, nausea, fever, back pain (mainly with streptokinase) (box 2).

F. Drug interactions :

Drug interactions with fibrinolytic drugs are relatively rare due to their short period of use.

♦**Increased risk of bleeding :**

- Non-steroidal anti-inflammatory drugs (NSAIDs), salicylates (aspirin) ;
- Antiplatelet agents ;
- Oral anticoagulants (antivitamin K, or AVK) ;
- Heparins (low molecular weight or unfractionated) ;
- Fondaparinux ;

- GPIIb/IIIa receptor antagonists.
- Increased risk of anaphylactic reaction: **conversion enzyme inhibitor.**
- **Physico-chemical incompatibility** (risk of precipitation) between :
- Alteplase and bivalirudin ;
- Reteplase and heparin.

2. Therapeutic strategies

The choice of fibrinolytic agent depends on an assessment of the benefit-risk ratio.

♦ **In cases of MI,** a reperfusion strategy should be undertaken as soon as possible. Fibrinolytics are generally required if access to a coronary angiography tray is not an option. In this case, tenecteplase is the most frequently used compound, as it has the advantage of being administered as a single bolus and has better haemorrhagic tolerance than reteplase. The high antigenicity of streptokinase greatly limits its use.

3. Associated advice :

4. Fibrinogen :

Fibrinogen concentration is highest in the morning, and lowest between 4pm and midnight, which may explain the morning hypercoagulability. This endogenous physiological property explains why fibrinolytics are more effective in the late evening. However, although this specificity should be taken into account, fibrinolytics are emergency treatments. It is impossible to decide when they should be administered, but it should be as early as possible.

5. After an acute coronary event :

A certain amount of advice and hygiene and dietary rules should be given to patients on discharge from hospital, as part of a progressive cardiovascular rehabilitation programme:

- smoking cessation (nicotine substitutes) ;
- a balanced diet (dietician);
- regular physical activity (about 30 minutes' walking a day);
- stress management.

6. After ST+ ACS :

It is important to ensure that patients are under the care of a general practitioner and a cardiologist, and that they are complying with their BASIC treatment (beta blocker, antiaggregant, statin, ACE inhibitor, control of risk factors) to prevent the risk of recurrence.

7. Procedure in the event of haemorrhage :

In the event of severe haemorrhage following the use of a 3rd generation fibrinolyte combined with heparin, protamine sulphate may be used. If this fails, transfusions of blood products or even the use of tranexamic acid (Exacyl®) may be necessary.

8. Treatment of anaphylactoid reactions :

Injection of fibrinolytics may cause exceptional anaphylactoid reactions. For this reason, it is recommended that treatment be stopped immediately if the first signs are detected. This may be followed by administration of antihistamines or corticoids, or even adrenaline.

1. Treatments associated with fibrinolytics :

1. Antiplatelet therapy :

Platelet activation and aggregation play a fundamental role in the formation of arterial thrombosis and are therefore major therapeutic targets in the management of acute coronary syndrome.

Platelets can be inhibited by three classes of drugs, each with a different mechanism of action:

- Aspirin (acetylsalicylic acid): by irreversibly inhibiting type 1 cyclooxygenase (COX-1), it blocks the formation of thromboxane A2 and reduces platelet aggregation.
- P2Y12 receptor inhibitors: this receptor is coupled to G proteins and is located on the surface of platelets. Its activation by adenosine diphosphate leads to the aggregation of blood platelets.

There are currently three inhibitors of this receptor: clopidogrel (CURE study [278], which has long been the reference P2Y12 inhibitor), prasugrel (TRITON TIMI 38 study [279] and ticagrelor (PLATO study [280]).

- GpIIb/IIIa antis: these involve blocking glycoprotein IIb/IIIa, the platelet receptor for fibrinogen. Three compounds are currently available: abciximab (grade A), eptifibatide (grade B) and tirofiban (grade B). They are indicated mainly in subjects at high risk of complications (infarction, death) for whom an early angioplasty strategy has been chosen.

2. Anticoagulant treatment :

Anticoagulants are used to inhibit thrombin generation and/or activity, thereby reducing thrombus-related events.

Various studies have demonstrated a synergistic anti-ischaemic effect when combined with antiplatelet therapy. Their use is therefore recommended in ACS (Grade I-A/C).

Numerous anticoagulants have been studied and act at different levels of the coagulation cascade:

- Unfractionated heparin (UFH)

It is a mixture of polysaccharide molecules with a molecular weight of 2,000 to 30,000 Da. A third of these molecules contain a polysaccharide sequence which binds to anti-thrombin, thereby increasing Tinhibition of factor Xa. The major disadvantage of TI INF lies in its narrow therapeutic range, requiring repeated measurements of Activated Cephalin Time to adapt the treatment dosage. TI INF was originally developed as a therapy, but is currently only used as a third-line treatment (Grade I-C).

- **Low molecular weight heparin (LMWH) :**

MPBIs have a lower molecular weight of between 2,000 and 10,000 Da and have anti-Ila and anti-Xa activity. Their major advantage is that they can be administered subcutaneously. However, they are partially contraindicated in cases of renal insufficiency (creatinine clearance < 30 ml/min). Enoxaparin is the best-studied compound and the most frequently used in clinical practice.

- **Fondaparinux :**

The only selective factor Xa inhibitor currently available, this pentasaccharide binds reversibly to Tanti-thrombin, catalysing its factorXa inhibitory action and thus preventing thrombin formation. Fondaparinux (2.5mg/d) has the most favourable efficacy/safety profile in non-ST+ ACS (Grade I-A), but is not recommended in ST+ ACS treated with primary angioplasty (Grade III-B).

- **Valirudine :**

This molecule binds directly to thrombin (factor IIa) and thus inhibits the transformation of fibrinogen into fibrin. Bivalirudin is the first-line treatment for primary angioplasty (Grade I-B).

3. Anti-ischaemic treatments :

Anti-ischaemic treatments can either reduce myocardial oxygen requirements (by reducing heart rate, blood pressure, pre-loading, etc.) or increase oxygen supply, notably by coronary vasodilatation.

a. Nitro derivatives :

Little studied in the acute phase of ACS, their use is mainly based on pathophysiological considerations and clinical experience.

Their benefits lie in a peripheral and coronary vasodilator effect that reduces myocardial oxygen consumption (by reducing pre-load and end-diastolic volume) and increases supply (coronary dilatation resulting in improved coronary flow).

They are the most frequently used:

- Symptomatic (stabilised angina pain)
- as a treatment for Acute Pulmonary Edema (APO)
- for diagnostic purposes ("trinitrin test")

b. Beta-blockers :

Beta-blockers competitively inhibit the myocardial faction of catecholamines, thereby reducing heart rate, blood pressure and contractility. Their efficacy in the acute phase is less established, as shown by a recent meta-analysis [281]. However, they are still indicated for intravenous treatment in the acute phase of ACS, in cases of arterial hypertension or tachycardia in the absence of heart failure (Killip > III) (Grade IIaB/C).

c. Calcium channel blockers :

The level of evidence for calcium antagonists is also low. What they all have in common is a vasodilatory effect, particularly in the coronary arteries, and in some cases an action on atrioventricular conduction and heart rate.

They have only a limited indication, often as a second-line treatment for ACS.

4. Complementary treatments :

a. Analgesic treatment *:*

Intravenous morphine titration is indicated for the relief of chest pain in ST+ ACS (Grade I-C).

Its main side effects are nausea and vomiting, hypotension with bradycardia and a risk of respiratory depression.

b. Treatment of anxiety

Angina pain is very anxiety-provoking, so the first effective measure is to reassure the patient by explaining the steps involved in the treatment (particularly in the case of angioplasty). Morphine is often enough to relax patients and relieve their pain. In the event of persistent anxiety, anxiolytic treatment with a benzodiazepine (e.g. dipotassium clorazepate) should be considered (Grade I-C).

c. Oxygen therapy :

Oxygen therapy is indicated in cases of hypoxia ($SpO2<95\%$), acute heart failure and/or respiratory difficulty (Grade I-C). The efficacy of systematic oxygen therapy (mask or goggle) is uncertain. Transcapillary oxygen therapy should, however, be monitored systematically.

d. Management of hyperglycaemia :

Hyperglycaemia is common in patients with acute coronary syndrome and is associated with increased mortality [282]. Treatment should aim to limit major hyperglycaemia (>2gl/l) while avoiding hypoglycaemia, which is a source of complications [283]. Blood glucose levels should be measured systematically during initial treatment, and repeated in cases of known diabetes or hyperglycaemia (Grade I-C). Its level should be maintained between 0.9 g/l and 2 g/l (Grade IIa-B).

J. Evolution :

- **KILLIP grade:** Killip grade III or IV is synonymous with poor haemodynamic status and is of a pejorative nature (LV dysfunction).
- **TIMI risk score:** This is designed to determine a patient's risk of mortality.

There is a linear relationship between the TIMI risk score and mortality.

Table VII: Calculation of the TIMI mortality risk score

Risk indicator	Points
Age greater than or equal to 75	**3**
Age between 65 and 74	**2**
History of diabetes, hypertension or angina	**1**
Systolic pressure below 100 mmHg	**3**
Heart rate > 100 beats per minute	**2**
KILLIP grade higher than I	**2**
Weight under 67 Kg	**1**
Previous location of STEMI	**1**
Reperfusion time greater than 4 hours	**1**
Possible total per patient	**14**

K. Management of initial complications

The initial complications of MI are discussed, taking into account their frequency and severity. Their specific treatment must be combined with correction of the factors contributing to them, in particular dyskalaemia and hypoxia.

1. Bradycardia

Several treatment options can be considered:

- Electrocardioscopic monitoring alone, on well-tolerated bradycardia and without risk of asystole ;
- External electrosystolic training, indicated for any bradycardia symptomatic with haemodynamic intolerance, usually related to high-grade AVB. It is also indicated if there is a risk of asystole occurring or if treatment with atropine is ineffective;
- Pharmacological treatment: in the absence of a reversible cause, atropine is the best choice.

the treatment of choice for any acute symptomatic bradycardiaê. Isoprenaline is not recommended and adrenaline should only be used as a last resort.

2. Tachycardias

Several treatment options can be considered (Grade B):

- Patients in circulatory arrest require an immediate asynchronous external electric shock (EEC) without sedation and medical cardiopulmonary resuscitation (CPR);

- Patients in haemodynamic failure (except in the context of sinus tachycardia) require immediate ERC, preferably synchronous;
- Conscious patients without haemodynamic failure, but with clinical signs of intolerance (chest pain or PAO) have their therapeutic strategy detailed in algorithm 4 ;
- Conscious patients without signs of clinical intolerance: sustained or polymorphic ventricular tachycardias (VT) justify the administration of amiodarone and/or a CEE after sedation. In other cases, therapeutic abstention with continued monitoring alone is the rule.

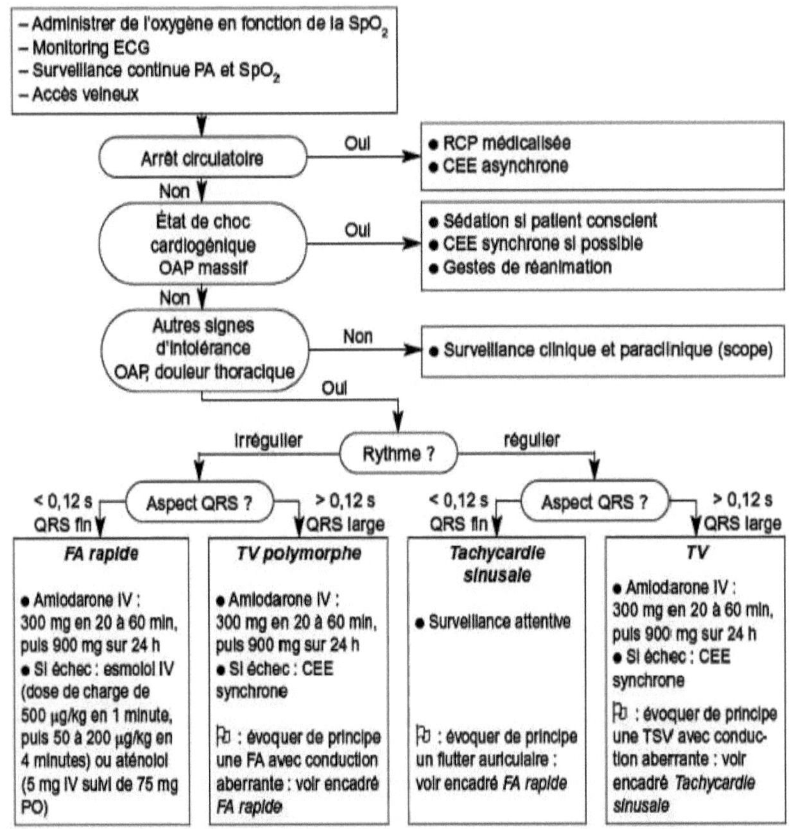

Algorithme 4. Therapeutic algorithm for tachycardia (heart rate > 100 bpm) in a patient with ACS, excluding vital failure and well tolerated tachycardia (after J.E. de la Coussaye et ***al).***

Figure 18: Treatment algorithm for tachycardia in patients with ACS

3. Circulatory arrest (CA) :

The ischaemic context of CA does not alter the general recommendations of the RCP.

CA occurs in a previously diagnosed MI: in a patient who has recovered spontaneous cardiac activity (RACS), the decision on the reperfusion strategy depends on whether there is rapid access to an operational DICT (diagnostic and interventional coronary angiography suite) (although no maximum time limit can be given in this context).

External cardiac massage does not contraindicate fibrinolysis.

- In a fibrinolytic patient, the occurrence of a CA may be a sign of coronary reperfusion. Prolonged resuscitation (60 to 90 minutes after injection of the fibrinolytic) is justified to promote its efficacy. In the absence of RACS, there is no scientific argument for recommending or prohibiting fibrinolysis.
- CA is the first manifestation of MI: there is no reason to recommend fibrinolysis in this context.

4. Cardiogenic shock

The therapeutic strategy is based on etiological treatment combined with symptomatic treatment, and includes early coronary desobstruction (grade B), preferably by angioplasty, combined with measures to reduce myocardial oxygen consumption (in particular analgesia and oxygen). Haemodynamic restoration is based on cautious vascular filling (in the absence of signs of left ventricular failure) with, if necessary, titrated use of catecholamines (dobutamine in 1st intention, noradrenaline in 2nd intention). Intra-aortic balloon counterpulsation (IABP) promotes initial stabilisation of patients in cardiogenic shock secondary to MI (Grade B).

5. Haemorrhagic events :

- Appropriate measures were taken on a case-by-case basis, ranging from simple monitoring with compression and readjustment of heparin doses to discontinuation of the thrombolytic, infusion of fibrinogen, measures to neutralise plasmin (Iniprol, Antagosan) or even blood transfusion and resuscitation in the event of cardiac failure.
- **Bleeding was classified according to severity using the TIMI bleeding scale:**

-Major bleeding: cerebral haemorrhage, or other bleeding accompanied by a drop in haemoglobin of more than 5 g/dl (retroperitoneal or occular).

-Minor bleeding: any bleeding accompanied by a drop in haemoglobin of between 3 and 5 g/dl.

-Minimal bleeding: bleeding resulting in a drop in haemoglobin of no more than 3 g/dl.

- Any change in cognitive function or even a sensory-motor deficit indicating intracranial bleeding. A brain scan is then essential to confirm the diagnosis and the topography of the lesion, and to guide the therapeutic approach which, in addition to stopping thrombolysis, may include intensive care and ventilatory assistance in the event of consciousness disorders.

6. Ischaemic recurrence :

It may be caused by extension into adjacent territories (painful recurrence with ST-T elevation in contiguous leads, whether or not associated with enzyme elevation). This complication threatens the short-term prognosis and requires urgent therapeutic measures such as haemodynamic support, or even coronary angiography and emergency revascularisation.

L. Evaluation of the management of ST+ ACS :

Learned societies and health authorities in France have selected a number of indicators to assess the quality of management of myocardial infarction in order to evaluate and improve professional practices with the aim of reducing the morbidity and mortality of this condition:

- **Rate of first-line use of the 15 call centre:** This is the percentage of patients with ST+ ACS or recent LBBB who have first-line use of the 15 call centre.

- **Median time between first medical contact and completion of qualifying ECG :** This is the time between the first medical contact (PCM: time of arrival, at the point of care, of a doctor able to perform the qualifying ECG) and the performance of the qualifying ECG. The qualifying ECG is defined as the tracing used to confirm the diagnosis of ST+ ACS. Current recommendations set this median time at 10 minutes.
- **Rate of implementation of an acute reperfusion strategy for ST+ ACS:** This is the percentage of patients with ST+ ACS or recent LBBB managed within 12 hours of the onset of symptoms and having
received a coronary reperfusion technique (thrombolysis or angioplasty).
- **Time to angioplasty for ST+ ACS:** This **is** the median time from the time of PCM to the time of angioplasty balloon inflation at the culprit lesion for patients with ST+ ACS or recent LBBB who underwent angioplasty within 12 hours of the onset of pain.

The current target is 90 minutes. In the case of very recent pain (<2 hours), this is reduced to 60 minutes.
- **Time to thrombolysis for ST+ ACS: This is the** median time between the PCM and the time of thrombolytic injection for patients with ST+ ACS or recent LBBB who have received fibrinolysis within 12 hours of the onset of pain. The optimal time is currently set at 30 minutes after the start of medical management.
- **Rate of direct referral to an Interventional Cardiology Centre (ICC) with 24-hour catheterisation:** This is the percentage of patients presenting with ST+ ACS or recent LBBB, managed within 12 hours of the onset of pain, and who have been referred directly to an interventional cardiology centre with 24-hour angioplasty facilities.
- **Rate of appropriate antiplatelet therapy: This** is the percentage of patients with ACS who received an appropriate prescription for antiplatelet therapy in the acute phase within 24 hours of the start of treatment.

NB: an appropriate prescription corresponds to a prescription in case of indication, or to an absence of prescription in case of contraindication, absence of indication or patient refusal.
- **Rate of appropriate anti-coagulant treatment:** This is the percentage of patients with ACS who were prescribed appropriate anticoagulation in the acute phase.
- **Morphine treatment rate for severe pain: This** is the percentage of patients with ST+ ACS or recent LBBB accompanied by persistent and severe chest pain (VAS > 60min or EN > 6) who received morphine treatment.
- **Rate of appropriate prescription of antiplatelet agents, statins, beta-blockers and converting enzyme inhibitors at discharge: This** is the percentage of patients with ACS who received an appropriate prescription of antiplatelet agents, statins, beta-blockers and converting enzyme inhibitors at discharge.

III/ The network :

A. A number of different actors and stakeholders are involved:

To ensure effective and sustainable health care, a well-functioning network is needed, made up of family doctors, emergency services and hospitals for triage, first aid and patient follow-up.

Apart from the patient, the various players and contributors to this network are the general practitioner and the paramedic on duty in the Mostaganem suburbs and in the triage department of the Mostaganem University Hospital, who are responsible for reception and initial medical contact, the ambulance driver or the emergency fireman, who are responsible for transporting or evacuating the patient, the cardiologist (state or private) or resuscitator who confirms the diagnosis, eliminates contraindications and performs thrombolysis, and the state or private interventional cardiologist who performs rescue angioplasty if thrombolysis fails.

B. Organisation:

ACS is a serious condition that can be life-threatening if diagnosis and treatment are delayed. In order to ensure effective and sustainable health care, a properly functioning network is needed for triage, first aid and patient follow-up.

In the absence of a catheterisation room close to Mostaganem, and with the aim of providing good emergency thrombolysis treatment for all our patients presenting with STEMI and ensuring their coronary reperfusion, the creation of a regional STEMI treatment network is absolutely urgent.

The organisation of this network is based on a working group made up of cardiologists, resuscitators and emergency doctors, which was set up to provide ongoing training for doctors at the region's various emergency points, as well as information for the general public. The aim of this network, organised in Mostaganem, is to thrombolise all patients who come to the UMC with an acute ST+ coronary syndrome within a maximum of 12 hours. The organisation of emergency care and transport for STEMI patients aims to achieve reperfusion within the first few hours.

The current configuration of the Wilaya of Mostaganem, with its ten (10) Dairas and thirty-two (32) communes, is the result of the third and last division dating from 1984 (law n°84-09 of 04/02/1984 and subsequent texts).

An analysis of the literature guided us in the choice of drugs to be used for coronary reperfusion by thrombolysis. In Mostaganem, thrombolysis is based in particular on the two fibrinolytics available at our level, Alteplase and Tenecteplase, with adjuvant treatment based on anticoagulation with low molecular weight heparins (EXTRACT TIMI 25 study) and double platelet anti-aggregation with aspirin and Clopidogrel (validated by the CLARITY-TIMI 28 trial).

For the time being, thrombolysis is still administered in hospitals, with the prospect of introducing it in pre-hospital settings.

C. Strategy:

The choice of our reperfusion strategy was based on current recommendations, in particular the French Consensus Conference and the 2017 ESC (European Society of Cardiology) recommendations, which distinguish between early (time between pain and qualifying ECG less than 3 hours) and late (time > 3 hours) ST+ acute coronary syndromes.

In our network, based on national and international recommendations, any patient in our region presenting with an acute ST+ coronary syndrome is referred to the UMC department in Mostaganem for rapid care by an emergency doctor and a resuscitation doctor available 24/7 in order to validate the STEMI diagnosis either directly or via telemedicine by a cardiologist available in our network, and to admit them to the thrombolysis unit as part of the activity of the UMC resuscitation unit.

We have also established a database of communes, with a transport time for each commune that must be respected in order to evacuate any STEMI patient, and a precise strategy defined beforehand in order to save on pre-hospital time.

Primary angioplasty is certainly the strategy of choice in the management of STEMI, but it is lacking in our region of Mostaganem because it requires a catheterisation room and a trained team available 24/7, and it must be performed within 120 minutes of the qualifying ECG.

D. Assessment :

The largest possible number of patients can only be cared for as part of an organised network, involving the UMC, the cardiology department with or without a catheterisation room, the EMS, GPs and general cardiologists. This network is centred around the UMC thrombolysis unit, and the pathway has been defined to reduce treatment delays. The private cardiology clinics in Mostaganem are fully integrated into this thrombolysis network. Protocols have been written and validated by emergency physicians and cardiologists. Information about the network was provided to healthcare professionals and individuals after it was set up.

Our procedure is regional, with primary admission always based in Mostaganem. The transport distance based on the communes distinguishes 2 groups (40/90 min).

The network is being evaluated by means of a register, in order to assess the quality of our practices, evaluate the management of ST+ acute coronary syndrome in terms of reperfusion strategy and patient referral, and finally improve this management by optimising delays and treatments.

Since the registry was set up on 1 December 2018, more than 300 STEMIs have been thrombolysed with a completeness rate close to 95%.

E. Mostaganem region:

1. Presentation:

The Wilaya of Mostaganem is a wilaya in Algeria in North Africa. It has **806408 inhabitants** over an area of 2,269 km[1] [2] . The population density of the Wilaya of Mostaganem is therefore 355.4 inhabitants per km2.

The current configuration of the Wilaya, with its ten (10) Dairas and thirty-two (32) communes, is the result of the third and last division dating from 1984 (law n°84-09 of 04/02/1984 and subsequent texts).

1. Daïra of Achaacha - 2. Daïra of Ain Nouissi - 3. Daïra of Ain Tadles - 4. Daïra of Bouguirat - 5. Daïra of Hassi Mameche - 6. Daïra of Khiredine - 7. Daïra of Mesra - 8. Daïra of Mostaganem - 9. Daïra of Sidi Ali - 10. Daïra of Sidi Lakhdar

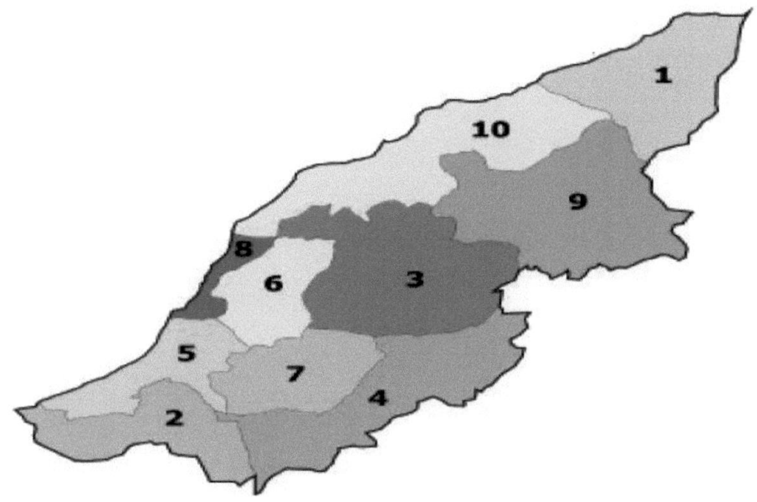

Figure 19: Location of the Jairas in the wilaya of Mostaganem

Like the rest of the Algerian population, the wilaya's population is young, with almost 39% under the age of 20. The 20 to 59 age group accounts for more than half of the wilaya's population. As a corollary, the population aged 60 and over is very small, accounting for just under 7% of the commune's total population. However, there has been a decline in the birth rate since the late 1980s.

The wilaya of Mostaganem includes a university hospital (CHU), 6 general hospitals (in Mostaganem, Sidi Ali, Aïn Tedles, Mesra, Bouguirat and Achaacha) and 2 specialist hospitals (EHS) (EHS Mère-Enfants Lala Kheira, EHS psychiatrique). The wilaya also has 6 local health establishments providing primary care (EPSP), with 28 polyclinics and 162 treatment rooms spread throughout the wilaya .[22]

2. Distance and route of the various dairas in the wilaya of Mostaganem:

Table VIII: Distance of dairas from the main town (wilaya of Mostaganem)

Dai'ra	Communes	Distance	Course
Mostaganem	Mostaganem		
Achaacha	Achaacha- Nekmaria- Khadra-OuledBoughalem	77 Km	1h20 Min
Ain Nouissy	Ain Nouissy - Fornaka- El Haciane	19 Km	28 MIN
Ain Tadeles	Ain Tadeles - Sour - Sidi Belatar - Oued El Kheir	22 Km	29 Min
Bouguirat	Bouguirat- Sirat- Safsaf-Souaflias	37 Km	44 Min
Hassi Mameche	Hassi Mameche - Stidia - Mazagran	11 Km	19 Min
Kheir Eddine	Kheir Eddine - Sayada- Ain Boudinar	10 Km	16 minutes
Mesra	Mesra- Mansourah- Touahria-Ain Sidi Cherif	17 Km	22 minutes
Sidi Ali	Sidi Ali - Tazgait- Ouled Maalah	46 kM	50 Min
Sidi Lakhdar	Sidi Lakhdar - Hadjadj - Ben Abdelmalek Ramdane	48.6 km	

3. Health infrastructure (wilaya of Mostaganem) in 2019:

Table IX: Health infrastructure (wilaya of Mostaganem) in 2019

HEALTH FACILITIES IN THE WILAYA

INFRASTRUCTURES SANITAIRES DE LA WILAYA			
Désignation	Nbre	Ratio Wilaya	Ratio national
Hôpitaux	08	1,62/ 1000 Hab	2 ‰ Public
Nbre de lits	1313	1,2 ‰	2 ‰
Polycliniques	28	1/29935 hab	1/25000 hab.
Salles de soins	162	1/5498 hab	1/5000 hab.

Figure 20: Location of the various hospitals in the wilaya of Mostaganem

F. Pathway for ST+ coronary patients less than 12 hours old in the Mostaganem region:

1. Recommended management pathway for ST+ ACS according to ESC 2017:

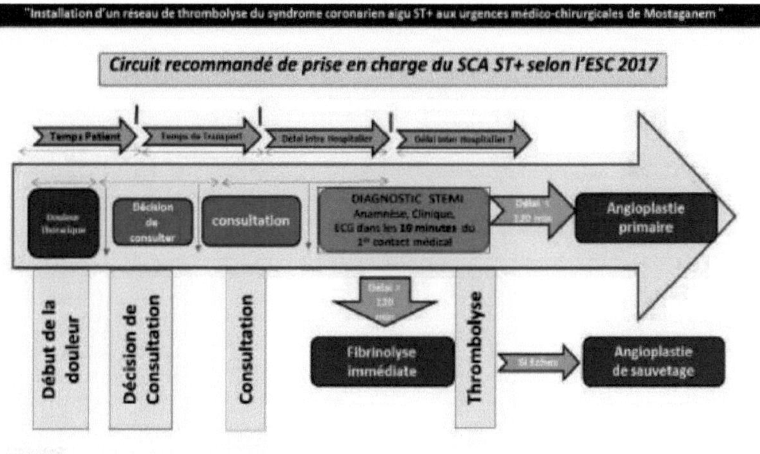

Figure 21: Recommended STEMI management pathway according to the ESC 2017

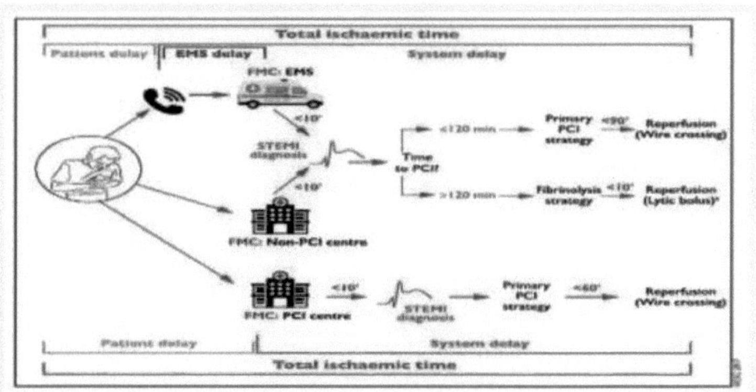

Figure 22: STEMI reperfusion strategy algorithm according to ESC 2017 recommendations.

2. STEMI circuit in Mostaganem before installation of the network :

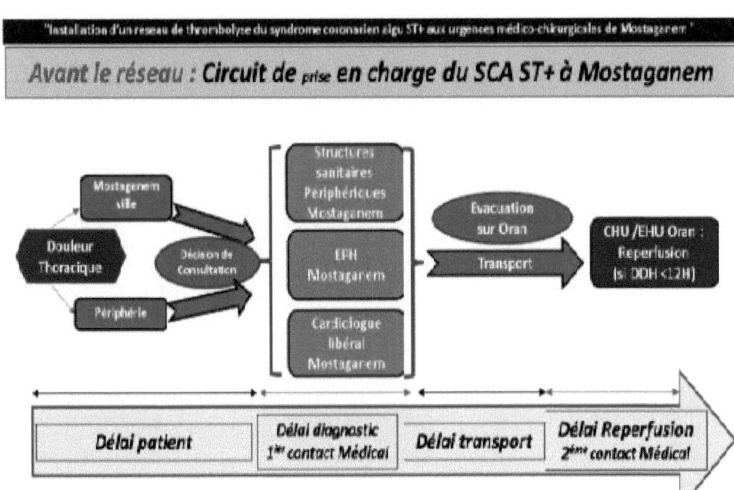

Figure 23: STEMI management circuit in Mostaganem before the installation of the network

c- STEMI circuit in Mostaganem after network installation :

Д with *the network:* ***ST+ ACS management circuit in Mostaganem***

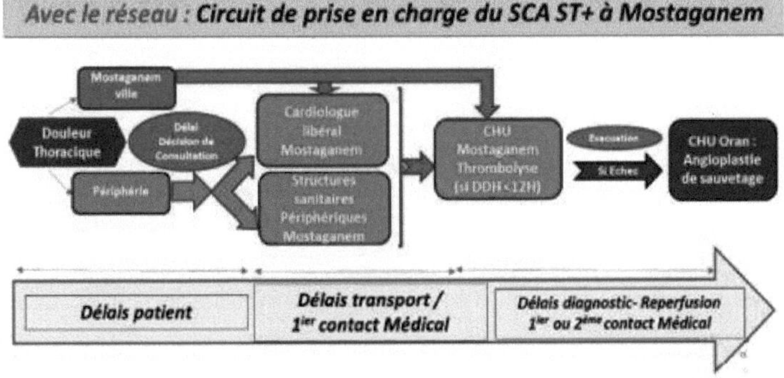

Figure 24: STEMI management circuit after installation of the network in Mostaganem

CHAPTER III

PRACTICAL STUDY

1. Study methodology :

1. Research hypothesis(es) :

> To propose a thrombolysis strategy adapted to our region by creating a thrombolysis unit and managing an effective network.
> Reducing STEMI morbidity and mortality in the wilaya of Mostaganem.
> Training and assisting emergency and EMS doctors in ECG reading and fibrinolysis.
> Request the equipment and medicines needed to deal with this vital emergency.
> Raise awareness of this emergency and its risk factors among the general public.
> Identify factors predictive of thrombolysis complications.
> Limit the time taken between the pain clinic, the diagnosis of STEMI and thrombolysis.

2. Aims of the study :

a. Main objective:

- Improving the management of STEMI patients by setting up a thrombolysis network at Mostaganem University Hospital's emergency departments.

b. Secondary objectives:

- Shorten the time between the clinical presentation of chest pain, the diagnosis of STEMI and thrombolysis.
- Identify factors predictive of thrombolysis complications.
- Propose a thrombolysis strategy adapted to our region.

I. Aims and impact of the study :

a. Educational :

- Training general practitioners in the investigation, diagnosis and management of acute coronary syndrome.

b. Scientific :

- Evaluation of a new approach to the management of coronary patients.

c. Public health :

- Propose a thrombolysis strategy similar to ours to other regions of the country that lack a catheterisation room, such as Mostaganem.

d. Socio-economic :

- Reduce complications and hospital stays, thereby saving human and financial resources.

II. Study protocol :

A. Materials and methods :

1. Type of study :

This is a prospective, single-centre, observational and descriptive study targeting patients hospitalised for STEMI with a delay between the onset of pain and hospitalisation of < 12 hours.

2. Study duration and link :

The duration of the study is estimated at 4 years in Mostaganem's emergency department.

3. Study population :

a. Inclusion criteria :

- All patients presenting with STEMI go to Mostaganem emergency department
- Women and men
- Aged between 18 and 75.
- The inclusion period: 2019- 2022.

b. Non-inclusion criteria :

- Patients presenting with chest pain but without ST+.
- Patients who have been progressing for more than 12 hours.
- Patients aged over 75.

4. Exclusion criteria :

These are the criteria that contraindicate thrombolytic treatment and therefore exclude patients from the protocol:

- Haemorrhagic stroke of any age
- Ischaemic stroke less than three months old
- Known intracranial neoplasia
- Known cerebrovascular malformation
- Suspicion of aortic dissection
- Active haemorrhage or haemorrhagic disease
- Major closed head or facial trauma less than three months old
- Poorly controlled hypertension
- Prolonged cardiovascular resuscitation (more than 10 minutes)
- Recent major surgery (less than 3 weeks)
- Recent internal haemorrhage (less than one month)
- Puncture of non-compressible vessels

-Pregnancy

- Active peptic ulcer
- Current anticoagulant treatment

5. Judging criteria :

The criteria for successful thrombolysis are clinical and electrical:

- The disappearance of pain.
- Reduction of the overshoot by at least 50%.
- The onset of an accelerated idioventricular rhythm (AIVR).

6. Minimum sample size :

- The minimum sample size will be 200 patients.
- It is calculated using the following formula: $\mathbf{N= z^2 . p. (1-p) / i^2}$

n = minimum sample size to obtain results. z = 95% confidence level (standard value of 1.96) p = estimated prevalence of STEMI (15%).

i = precision of the study. Margin of error at 5% (standard value of 0.05).

- So for an event that is 15% likely, with a confidence level of 95% and a margin of error of 5%, the minimum sample size will be: N=1.962x0.15x0.85/0.052 = 195.9.
- Our minimum sample size is 196 patients, but this can be increased to 200 patients.

111. Conduct of the study :

We conducted a prospective study in the emergency department of Mostaganem

University Hospital over a four-year period, from December 2018 to November 2022, during which we thrombolysed all the patients who presented to us in emergency with a STEMI less than 12 hours old, i.e. 295 patients in all.

The data was collected by the emergency doctor or by the specialist (cardiologist or intensive care unit) on duty, either immediately on admission or after a delay due to the urgent nature of the treatment.

Our management was carried out in accordance with international recommendations for STEMI, after informing the patient of the risk and vital interest of thrombolysis, as well as of the risks and benefits of thrombolysis.

objectives of the study.

On admission, questioning of the patient or relatives should seek to identify :

A. Consultation method and patient pathway :

- Arrived directly from his home at A&E,
- Referred by a doctor (public or private),
- Or medically evacuated from another hospital.

B. Collection of patient data :

- Age and gender.
- Weight, height, mass index [BMI=weight (KG)/height2 (m2)].
- Address.
- Level of education.
- Social status.
- Employment.

C. Full clinical assessment :

1. Research into cardiovascular risk factors :

- **Age** greater than or equal to 50 for men and 60 for women
- Active smoking, passive smoking or smoking cessation in less than three years.
- **Arterial hypertension:** in line with European recommendations.

Table X: Hypertension according to European recommendations

Category	**PAS (mmHG)**		**PAD (mm HG)**
Mild hypertension	**140-159**	**and/or**	**90-99**
HTA limit	**140-149**	**and/or**	**90-94**
Moderate hypertension	**160-179**	**and/or**	**100-109**
Severe hypertension	**> 180**	**and/or**	**> 110**
Isolated systolic hypertension	**> 140**	**And**	**< 90**
Borderline systolic hypertension	**140-149**	**And**	**< 90**

- **Diabetes: defined according to AFSSAPS standards as** a glucose level > 1.26 g/l.
- **History of coronary disease**: angina, MI, angioplasty and bypass surgery.
- **Dyslipidaemia**: which includes, alone or in combination :

1. LDL-cholesterol level >1.6g/l.
2. HDL-c levels <0.40g/l
3. Triglyceride levels >1.5g/l

4. Total cholesterol > 2g/l

- **Coronary heredity.**
- **Overweight**: according to the criteria of the International Obesity Task Force.

Table XI: Overweight according to the criteria of the International Obesity Task Force.

Classification	BMI
Overweight	**25,0-29,9**
Obesity	**> 30,0**

- **Menopause.**
- **Stress.**

2. Personal history :

- Consumption of alcohol, cannabis and other drugs.
- A history of stroke or transient ischaemic attack.
- A history of peripheral arterial disease or thromboembolic venous disease.
- History of chronic kidney disease: CKD.
- History of chronic respiratory disease: Asthma and COPD.
- History of thyroid disease: hyper- and hypothyroidism.
- History of gastric pathology: ulcers and gastritis.
- A history of complete atrial fibrillation arrhythmia (CAFA).
- History of haemorrhage.
- Treatment in progress.
- A recent intramuscular injection.

3. Clinical diagnosis on admission :

Looking for :

- Chest pain (angina): type, intensity, timing and mode of onset.
- Epigastric pain (possibly equivalent to angina).
- Dyspnoea and functional stage according to the N.Y.H.A. classification.
- KILLIP classification
- Heart rate, blood pressure.
- Pulmonary rales and signs of right heart failure.
- Nitrate derivative test: see the evolution of angina on 5mg Risordan sublingual.
- **The different deadlines :**
- Time of onset of chest pain.
- Time of first consultation.
- Time of first specialist medical contact.
- Time of arrival of the patient "night or day".

- **Killip classification**: assesses the patient's haemodynamic status in four grades:

- Grade 1: hemodynamically stable.
- Grade 2: borderline haemodynamic.
- Grades 3 and 4: a poor haemodynamic state or even cardiogenic shock.

D. Electrocardiogram :

17 derivations to specify :

- Heart rate.
- Heart rhythm: sinus, ACFA.
- Repolarisation disorders suggestive of ischaemic signs:

- o ST Segment: Sus shift (amplitude, territory).
- o T wave: subepicardial ischaemia (amplitude, territory).
- o Sequelae of necrosis (territory)
- o Conduction disorder (BBG, BBD, BAV)
- o Rhythm disorders (ESV, VT)

- ST-T elevation in millimetres (mm), the main inclusion criterion.

In all cases, the anamnesis and ECG were sufficient to make the diagnosis of a developing myocardial infarction and justify the initiation of fibrinolytic treatment.

E. Biological check-up :

Including as a matter of urgency :

- Blood glucose.
- Blood grouping.
- Renal function: urea, creatinine
- Blood count formula: haemoglobin, platelets, white blood cells.
- Measurement of ultrasensitive troponins (Tus).
- Haemostasis work-up: determination of prothrombin levels TP/INR and TCA
- HBA1c levels.
- Lipid balance

F. Therapeutic management :

Two treatment protocols are required:

- **Primary angioplasty (the gold standard)** is the safest and most effective technique, reopening the occluded artery in over 90% of cases, compared with only 60% for fibrinolysis.
- **Fibrinolysis** has the advantage of simplicity, as it can be performed anywhere in the country. It is most effective in the first 3 hours after the onset of symptoms. The risk of intracerebral haemorrhage (between 0.5 and 1%) is unavoidable, despite strict adherence to contraindications. Two fibrinolytics are recommended for repermeabilisation. These are Tenecteplase and Alteplase.
- The HAS experts recommend the preferential use of tenecteplase, a fibrin-specific product that can be used as a single intravenous bolus over approximately 10 seconds, with a short half-life, adaptable to the patient's weight and a dose not exceeding 10,000 IU, i.e. 50 mg of tenecteplase compared to Alteplase.

■ **Associated therapies :**

- **Heparin therapy:** Lovenox is the only LMWH offering a therapeutic alternative to unfractionated heparin in this indication. It is administered as 2 subcutaneous injections per day (12 hours apart), each containing 100UI Anti-XA/kg.
- **Platelet anti-aggregants**: 250 mg bolus followed by 160-300 mg/day orally.
- **Beta blockers**: Metoprolol (200 mG tab) and Atenolol (100 mg tab) were systematically prescribed in the absence of any alteration in haemodynamic status or contraindication (AVB, heart failure, bronchial asthma or heart rate below 50 Bpm). They help to attenuate the initial adrenergic reaction.
- **Converting enzyme inhibitors (CEIs)**: Captopril (25 MG tab) and Ramipril (1.25 and 5 Mg tabs) were prescribed initially as a test dose and then in increasing doses depending on blood pressure response. The usual contraindications were respected (renal

insufficiency, arterial hypotension, VD extension). ACE inhibitors, which reduce LV remodelling, were continued for a minimum of six to eight weeks, or even longer in cases of left ventricular dysfunction.

- **Analgesics:** such as Nefopam (ACUPAN) have been given in hyperalgesic forms, subject to contraindications (respiratory insufficiency).
- **Positive inotropes**: indicated in cardiogenic shock or right ventricular infarction. We used Dobutamine at 5 to 25 γ/Kg/min or Dopamine at 3 to 20γ/Kg/min.
- **Diuretics**: 20 mg Furosemide injection for left ventricular failure.
- **Statins**: prescribed unless contraindicated (intolerance, hepatic insufficiency, steatosis)

G. Echocardiography :

It was carried out within the first 48 hours of hospitalisation. This examination was carried out at the patient's bedside. It was used to analyse left ventricular function using Simpson's method (global LVEF).

A left ventricular ejection fraction of less than 40% is a sign of left ventricular dysfunction.

H. Evolution :

- **Short-term:** represents the patient's progress during thrombolysis and immediate post-thrombolysis up to forty-eight hours of hospitalisation in an intensive care unit (ICU).
- **Clinical course:** the intensity of chest pain is noted on arrival and reassessed after thrombolysis to look for any exacerbation or reduction. If the pain disappears within the first three hours of thrombolysis, this is an indication of coronary reperfusion, whereas if it persists, reperfusion has failed.
- **Haemodynamic status: the Killip grade** is reassessed in the ICU.
- **Electrocardiographic changes :**

The appearance of a necrotic "Q" wave or the onset of an accelerated idioventricular RIVA rhythm is noted. The ST segment is analysed after thrombolysis. A reduction of at least 50% in ST elevation within 3 hours of thrombolysis and the onset of an accelerated idioventricular rhythm (RIVA) are strongly suggestive of coronary recanalisation, while failure to reduce ST elevation suggests that the artery has remained occluded. The appearance of a Q wave is not necessarily synonymous with non-recanalisation.

- **Changes in echocardiographic parameters :**

A full cardiac ultrasound was performed in all patients at forty-eight hours in the event of a favourable outcome, at the end of the ICU stay.

- **Hospitalization in ICU :**

If the outcome is favourable, the patient is transferred to cardiology within 48 hours. If the outcome is unfavourable, the patient is kept under haemodynamic support.

If reperfusion failed, oral nitrates were continued and the patient was referred to Oran for coronary angiography.

Discharge from ICU and transfer to cardiology :

To be carried out on Day 2 if the initial course is favourable.

I. Complications :

- **Any deterioration in haemodynamic status**: this may be due to LV dysfunction, hypovolaemia, right ventricular (RV) extension or a rhythm or conduction disorder (ECG monitoring) requiring therapeutic adjustments and/or urgent measures (cardioversion, antiarrhythmics, atropine, electro systolic training, filling and positive inotropes).

- **Any external or occult bleeding**: which has been attributed to thrombolytic treatment; to do this, the puncture sites, oral cavity and mucocutaneous compartment are systematically examined for ecchymotic plaques.

J. Mortality :

Deaths are defined as cardiovascular when they are fatal heart attacks or deaths that are not due to a clearly non-cardiovascular cause, such as deaths from cancer, infection, trauma or suicide.

One-month mortality was considered in the following way: deaths before the 6^{e} hour of admission, deaths before the 24th hour and finally deaths occurring between one and thirty days.

The clinical characteristics of the patients who died were compared with those of the survivors.

K. End and end of the study :

A patient is considered to have completed the study at the time of death or when clinical observation has been completed at three months.

Patients may be withdrawn from the study due to an intercurrent event, death, a patient "lost to follow-up" or withdrawal of consent ("discharge against medical advice").

L. Presentation of results and statistical tests :

The results are expressed by age group: four age groups are defined: patients aged under 45, patients aged 45 to 54, patients aged 55 to 64 and patients aged 65 to 75.

Comparisons of quantitative variables were made using the F-test (analysis of variance) and Student's t-test.

Comparisons of categorical variables were made using the Chi-square test.

The data were **analysed** using SPSS 26 software.

M. Ethics :

The informed consent of all patients was considered necessary and was respected in this study (see Appendix 1).

IV. Results :

A. Before installing the network :

Before the installation of our network at the EPH of Mostaganem, and in the register of the cardiology department, during the first 11 months of 2018, we found only 10 thrombolysed patients of all ages, as shown in figures and tables 11, 12 and 13, mainly due to the lack of cardiologists in the hospital, and STEMI patients from the region were obliged to be evacuated urgently to Oran University Hospital, at the cost of being treated at very long intervals and subject to complications.

Table XII: Number of people thrombolysed before the network, by sex

Gender	Frequency	%
Men	6	60
Woman	4	40
Total	10	100

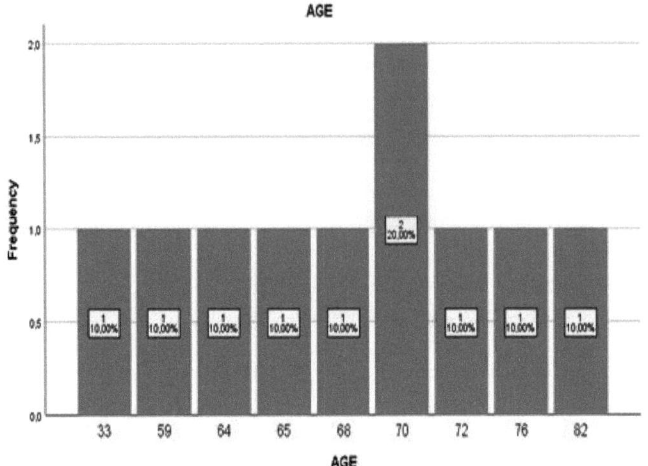

Figure 25: Age distribution of patients receiving thrombolysis in 2018 before the network was set up

Table XII: Number of people thrombolysed before the network, by sex and by age group

Sex / Age	<45 years	Age 55 to 64	65 to 75 years	> 75 yearsrc	TAL
Men	1	1	4	0	6
Woman	0	1	2	1	4
TOTAL	1	2	6	1	10

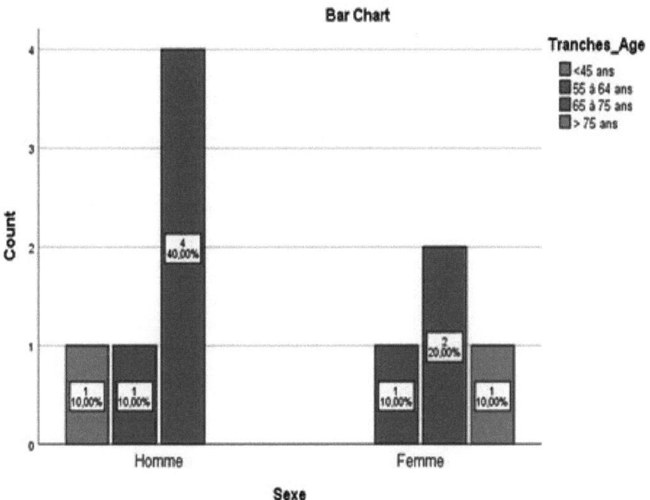

Figure 26: 2018 headcount by gender and age group

Table XIII: Fibrinolytics used by sex before the network

Sex / Age	TENECTEPLASE	ALTEPLASE	TOTAL
Men	**4**	**2**	**6**
Woman	**4**	**0**	**4**
TOTAL	**8**	**2**	**10**

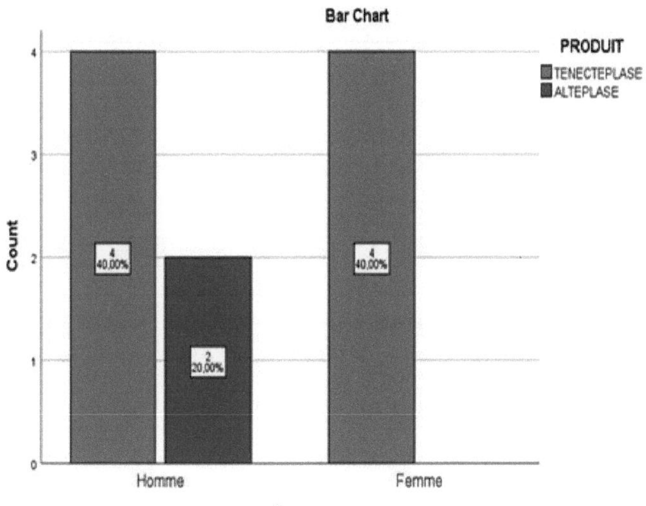

Figure 27: Number of patients treated with thrombolysis, by sex and by drug

B. After installing the network :

Our analysis concerns the data of 295 patients from the different regions of Mostaganem who were collated and thrombolysed during four consecutive years (36 months) for STEMI in the emergency department of Mostaganem between December 2018 and November 2022.

Table XIV: Number of STEMI patients thrombolysed per year after the network was set up

Year	Frequency	%
2019	**66**	**22,4**
2020	**72**	**24,4**
2021	**72**	**24,4**
2022	**85**	**28,8**
Total	**295**	**100,0**

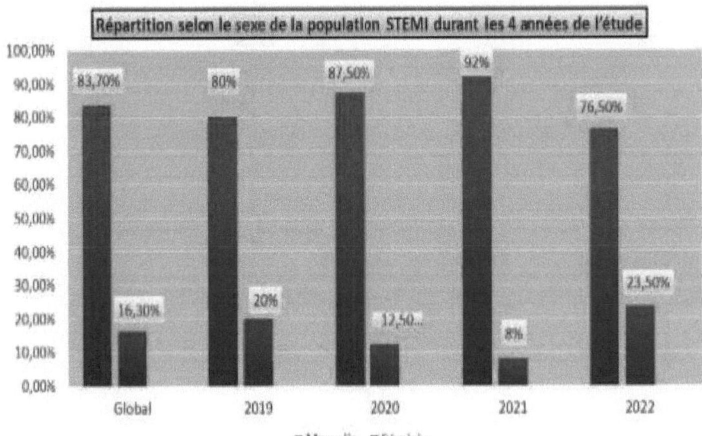

Figure 28: Number of thrombolysed patients in the study, by year and sex

2. Characteristics of the population :

a. Breakdown by gender :

Our population of 295 coronary patients comprised 247 men **(83.73%)** and 48 women **(16.3%)**. There was a clear male predominance, with a sex ratio of **5.1**, i.e. five men for every woman.

Table XV: Breakdown of STEMI population by gender

Workforce	Frequency	%	Total	Sex Ratio
Men	**247**	**83,73%**	**295**	5,14
Women	**48**	**16,3%**	**100%**	

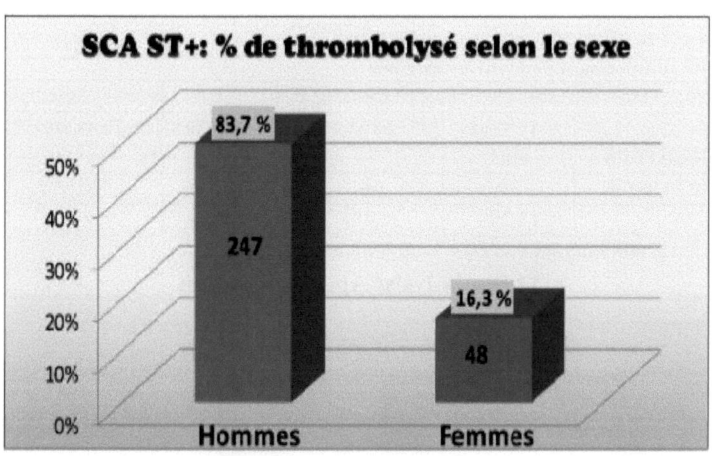

Figure 29: Number of thrombolysed patients in the study, by sex

b. Average age of the study population :

The average age of our total population is **57.85** (with extremes of 27 and 75).

Women were slightly older (mean age **62.13**) than men (mean age **57.02**) with a p = 0.002.

Table XVII: Average age of sample by sex and diabetes

	AVERAGE AGE	INTERVAL	WORKFORCE	P
SERIES	**57,85 ± 10,51**	**Age 27-75**	**295**	**-**
MEN	**57.02 ± 10,52**	**Age 27-75**	**247**	
WOMEN	**62,13 ± 9,47**	**Age 38-75**	**48**	**0,002**
DIABETICS	**62,26 ± 8,37**	**42-75 years**	**80**	
NON-DIABETICS	**56,21 ± 10,77**	**Age 27-75**	**215**	**0,0001**

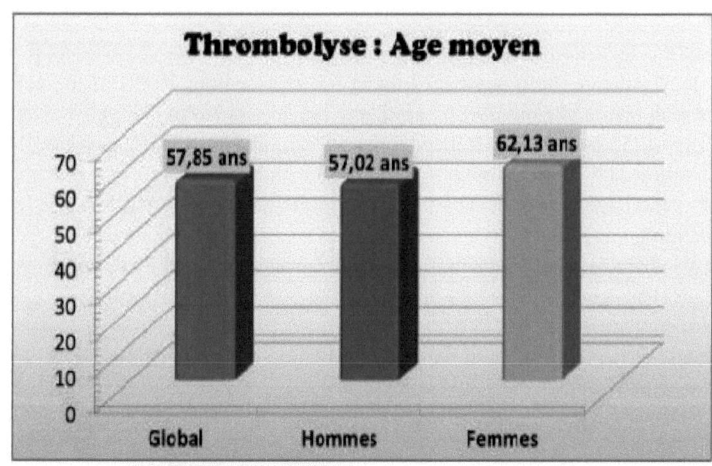

The patient on 4-)еилй 27 years **The patient on 4-ãgí!** 75. ansí

Figure 30: Number of thrombolysed patients by gender

c. Population distribution by age group :

Table XVIII: Distribution of workforce by age group

Distribution of workforce by class t				age	
AGE GROUPS	< 45 years old*	[45 -54 years] *	[55 -64 years] *	[65-75 years] *	Total
WORKFORCE	39	66	102	88	295
%	13,2%	22,4%	34,6%	29,8%	100%

** : $p<10^{-3}$ (significant difference)*

Thrombolysis: Age distribution

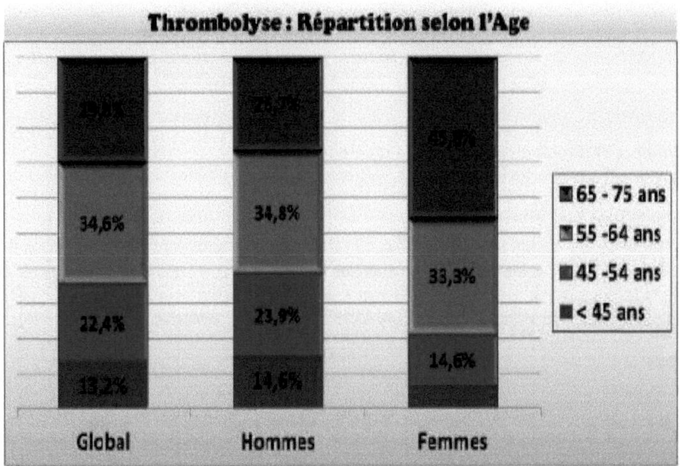

Figure 31: Breakdown of STEMI population by age group

Analysis of the age curve identified four age groups: < 45, [45-54], [55-64] and [65-75]. Two-thirds of patients (209) were aged under 64, representing 70.8% of the population, compared with one-third (86) (29.2%) aged between 65 and 75 (p=0.001).

The age group [55-64 years] was predominant in our study, with 102 patients (34.6% of the total), followed by the age group [65-75 years] with 88 patients (29.8% of the total).

The [< 45 years] age group remains the least represented with 39 patients (i.e. 13.2% of the population).

It should also be noted that 66 patients (22.4%) were aged between 45 and 54.

207 patients were aged under 65 on the day of the infarction, representing 70.17% of the population, compared with 88 (29.8%) aged between 65 and 75.

While almost a quarter (26.7%) of our male population is over 65, almost half (45.8%) of our female population is over 65 (p = 0.001).

Figure 21 shows the distribution of the occurrence of ST+ ACS according to the age and sex of the patients in our study. It shows a peak frequency between the **ages** of **55 and 64** for men and between **65 and 75** for women.

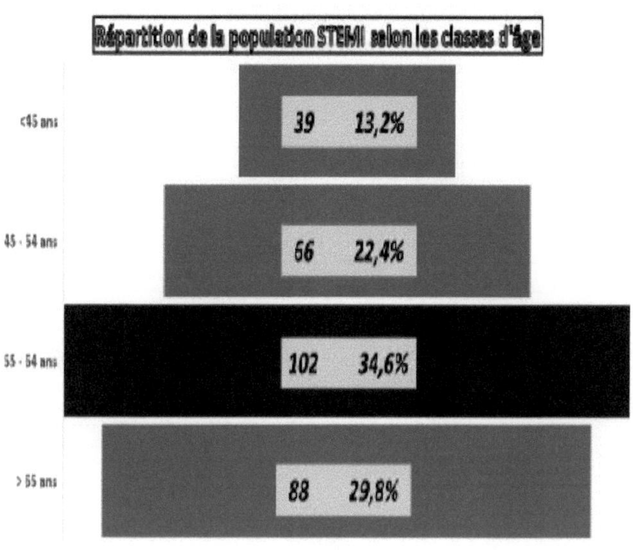

Age moyen = 57,85

Figure 32: Age curve for the STEMI population

Table XIX: Distribution of workforce by age group and gender

AGE GROUPS	< 45 years	[45 -54 years]	[55 -64 years]	[65-75 years]	P
Men	**14,6%**	**23,9%**	**34,8%**	**26,7%**	
Woman	**6,3%**	**14,6%**	**33,3%**	**45,8%**	0,035
Total	**13,2%**	**22,4%**	**34,6%**	**29,8%**	

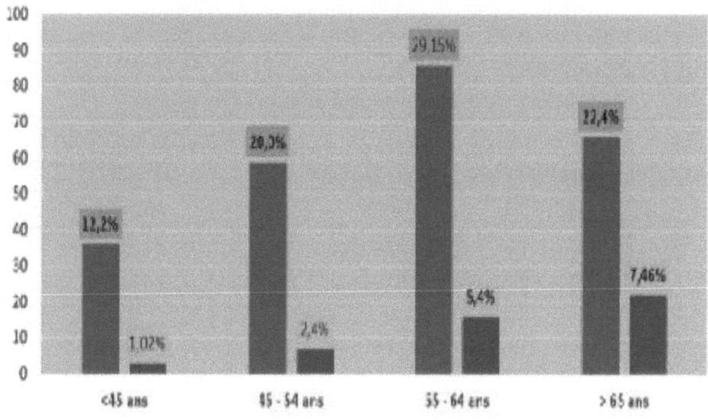

Figure 33: Distribution of the study population by gender and age group

d. *Socio-professional data*

In the course of our study, we interviewed patients or their carers to find out their socio-professional status. Working people and retired people represented **53.6% and 33.6%** of the population respectively. No patient was disabled or on long-term sick leave.
While 60.73% of the men in our sample population are in work, 64.58% of the women are homemakers.

Table XX: Professional status of the STEMI population by gender

PROFESSION	Assets	Retired	Unemployed	P
Men	60,7%	36,4%	2,8%	0,0001
Woman	16 ,7%	18,8%	64,6%	
Total	53,6%	33,6%	12,9%	

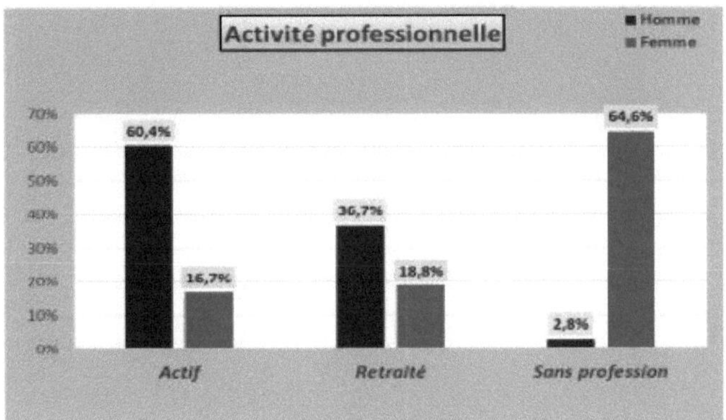

Figure 34: **Professional status of the STEMI population by gender**

2. *Medical data :*

a. Mode of admission and patient pathway (first medical contact) :

In the absence of a mobile call service in Mostaganem, 43.4% of the patients in our population went directly to emergency with their own means of transport, 39.7% of patients were evacuated by ambulance from the various duty points in the wilaya and the remaining 16.9% of patients were referred from a private general practice or a specialist cardiology or intensive care unit.

Table XXI: Mode of patient admission by gender

Admission procedure	Consultation	Evacuation	Orientation of a private practice	P
Men	43,7%	40,9%	15,4%	0,245
Woman	41,7%	33,3%	25,0%	
Total	43,4%	39,7%	16,9%	

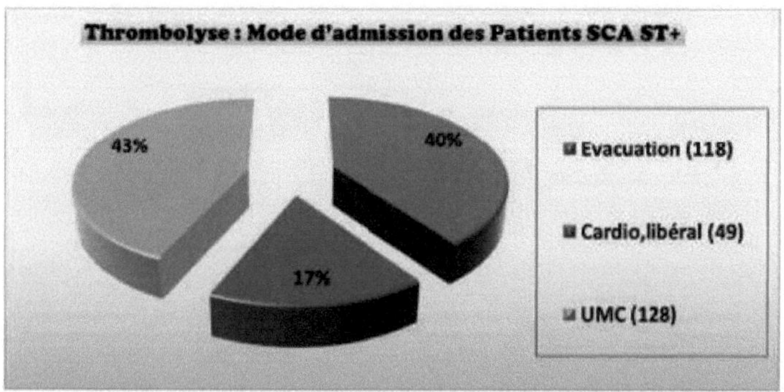

Figure 35: Overall mode of admission for STEMI patients

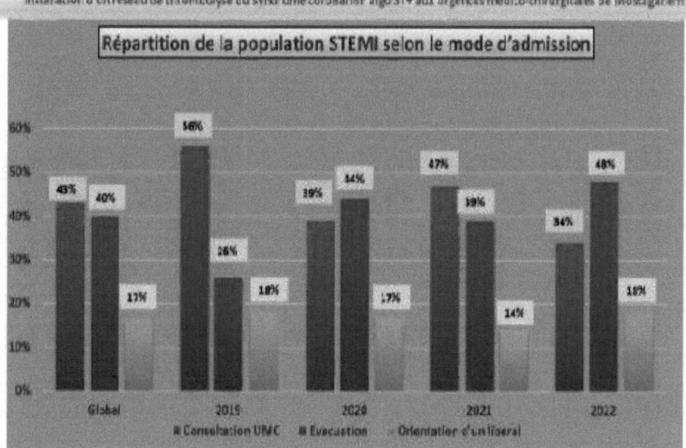

Figure 36: Mode of admission for STEMI patients by sex and year

e. Origin of STEMI patients :

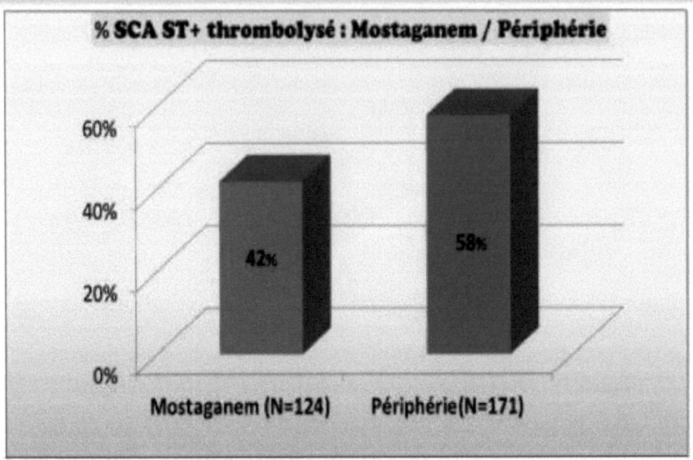

Figure 37: Origin of STEMI patients

Table XXII: Home address of STEMI patients by sex

Place of origin	Men	%	Woman	%	Total	%
Achaacha	6	2,43	0	0	6	2,0
Ain Nouissy	12	4,86	1	2,08	13	4,4
Ain Tadles	20	8,10	4	8,33	24	8,1
Bouguirat	27	10,93	2	4,17	29	9,8
Kheireddine	13	5,26	2	4,17	15	5,1
Mamache	20	8,10	7	14,58	27	9,2
Mesra	29	11,74	12	25	41	13,9
Mostaganem	99	40,08	19	39,58	118	40,0
Passenger	9	3,64	0	0	9	3,1
Sidi Ali	6	2,43	1	2,08	7	2,4
Sidi Lakhdar	6	2,43	0	0	6	2,0

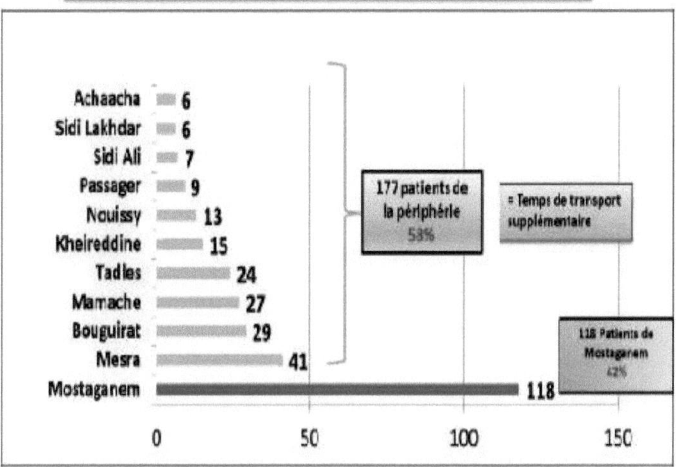

Figure 38: Breakdown of STEMI patients' home addresses

f. Time of day or night admission of STEMI patients

Patients were admitted three times more often during the day than at night (73.22% vs. 26.78%).

Table XXIII: Time of day/night admission of patients by gender

Time of admission/ Sex		8H- 20H	%	20H- 8H	%
MEN	**247**	**176**	**71,3%**	**71**	**28,7%**
WOMAN	**48**	**40**	**83,3%**	**8**	**16,7%**
TOTAL	**295**	**216**	**73,2%**	**79**	**26,8%**

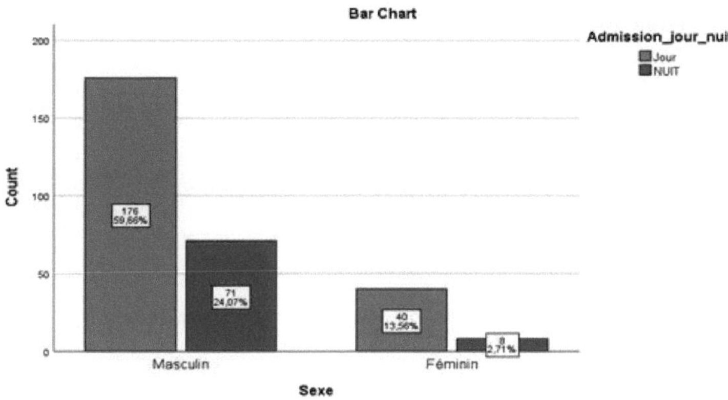

Figure 39: Breakdown of time of admission by gender

g. Timing of STEMI occurrence during the day :

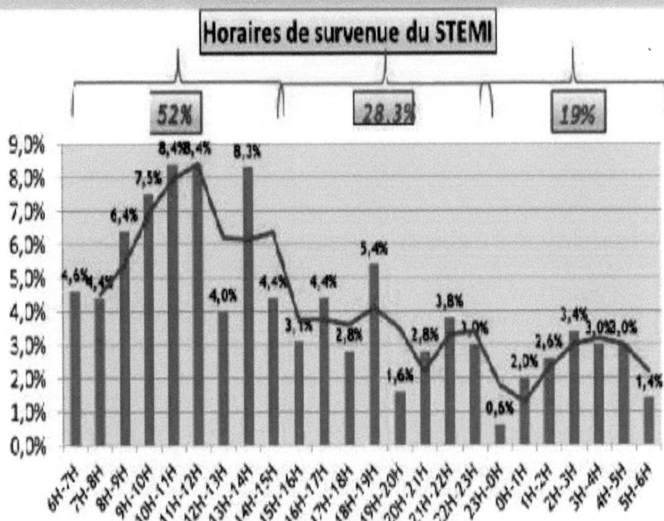

Figure 40: Time of day when STEMI occurs

h. **Timetable for thrombolysis of STEMI during the day :**

"Installation d'un réseau de thrombolyse du syndrome coronarien aigu ST+ aux urgences médico-chirurgicales de Mostaganem"

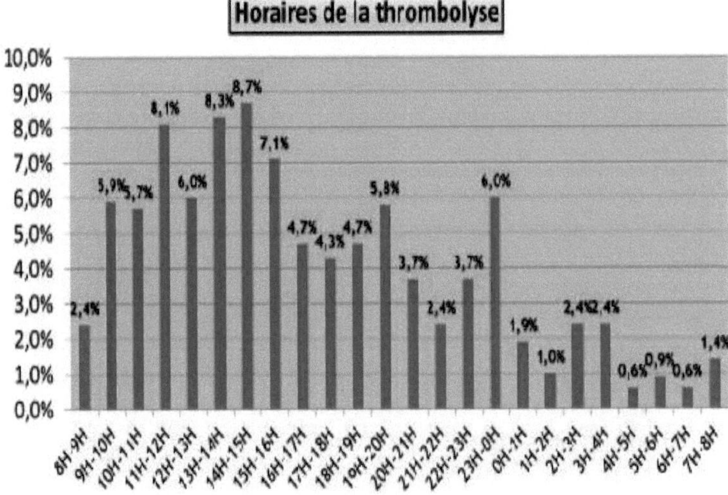

Figure 41: Timetable for thrombolysis of STEMI during the day

4. Admission deadline :

a. Average length of stay in hospital :

The average time taken to hospital in our study **was 3 hours 42 minutes +/- 2 hours 24 minutes** from the onset of symptoms, reflecting patients' lack of awareness of the importance of seeking medical attention early in the event of prolonged chest pain.

Depending on gender, the average thrombolysis time was longer for women than for men (4H50 min ± **2H47 min versus 3H40 min ± 2H16 min. p = 0.002).**

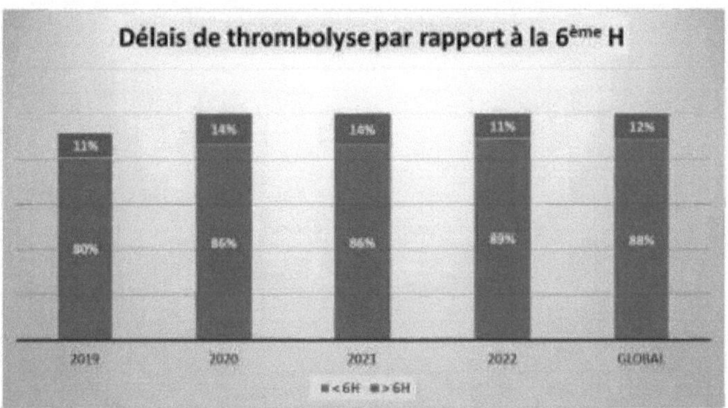

Figure 42: Thrombolysis time to 6ème hours

<u>Table XXIV: Average time taken by the various outpatient Jairas before arrival at the emergency department</u>

Average waiting times for various out-of-hospital dairas before emergencies			
Evacuated from	**Medium Course**	**N**	**Std. Deviation**
Achaacha	**5:10**	**6**	**2:21**
Ain Nouissy	**4:23**	**13**	**2:51**
Ain Tadles	**5:05**	**24**	**2:48**
Bouguirat	**5:12**	**29**	**2:17**
Kheireddine	**3:34**	**15**	**2:18**
Mamache	**2:22**	**27**	**1:36**
Mesra	**4:35**	**41**	**2:18**
Mostaganem	**2:46**	**118**	**1:56**
Passenger	**3:33**	**9**	**2:51**
Sidi Ali	**4:42**	**7**	**1:56**
Sidi Lakhdar	**6:10**	**6**	**2:46**
Total	**3:42**	**295**	**2:25**

b. General characteristics :

For the series as a whole, **94** patients **(31.86%)** were thrombolyzed before the 2nd hour, **157** patients **(53.22%)** before the 6th hour and **44 (14.92%)** after that. The earliest time for thrombolysis was 30 minutes after the onset of pain, and the longest was 12 hours.

Age groups: There was no significant difference between age groups in terms of thrombolysis time.

In relation to diabetes: 64 **diabetic** patients **(80%)** out of a total of 80 were thrombolyzed **before the 6th hour**, compared with 195 **non-diabetic patients (90.7%)** out of a total of 215. The difference was significant in favour of non-diabetic patients (p =0.017).

There was no significant difference between the daytime and night-time timing of thrombolysis (224 min versus 215 min with a p = 0.619).

Table XXV: Time to admission of STEMI patients by sex

Admission times by gender				
Admission deadline /Gender	**< 2H**	**2-6H**	**>6H**	P

Male	84(34%)	132(53,4%)	31(12,6%)	
Female	10(20,8%)	25(52%)	13(27%)	0,019
Total	94(32%)	157(53%)	44(15%)	

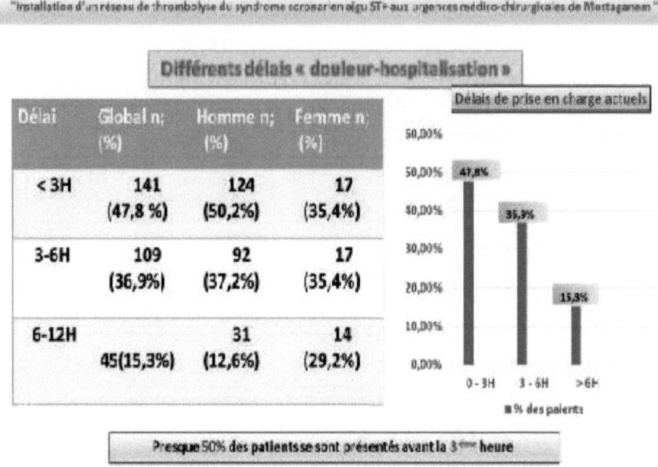

Délai	Global n; (%)	Homme n; (%)	Femme n; (%)
< 3H	141 (47,8 %)	124 (50,2%)	17 (35,4%)
3-6H	109 (36,9%)	92 (37,2%)	17 (35,4%)
6-12H	45(15,3%)	31 (12,6%)	14 (29,2%)

Figure 43: Different delays between pain and hospitalisation

The average time between the onset of pain and arrival at emergency or [pain-emergency] was 231 min ±. It varied from 30 min to 11H45min. It was **250 min** ± 176 min for **diabetic** patients compared with **211 min** ± 130 min for non-diabetic patients, with a significant difference (P = 0.042).

The [pain-emergency] time was significantly longer in **women** (**278 min** ± 175 min) than in men (**211 min** ± 136 min) (P=0.003).

The [pain-emergency] delay was significantly ($p < 10^{-3}$) longer for patients **evacuated** from the various health facilities (**270 min** ± 133 min) and even longer for patients consulting **a private doctor** (**327 min** ± 164 min) compared with a delay of (**137** min ± 91 min) for patients coming directly to emergency.

Table XXVI: Time to thrombolysis for patients with diabetes

Time limits for throm		bolyse by diabetes N=295			
	< 2H	] 2H-6H]	< H6	>6H	
SERIES 295	94 (32%)	157 (53%)	251 (85%)	44 (15%)	P
DIABETICS 80	28(9,5%)	36(12%)	64(21,7%)	16(5,4%)	0,006
NON-DIABETICS 215	66(22,4%)	121 (41%)	187(63,4%)	28 (9,5%)	

Différents délais /6ème Heure « douleur-thrombolyse »

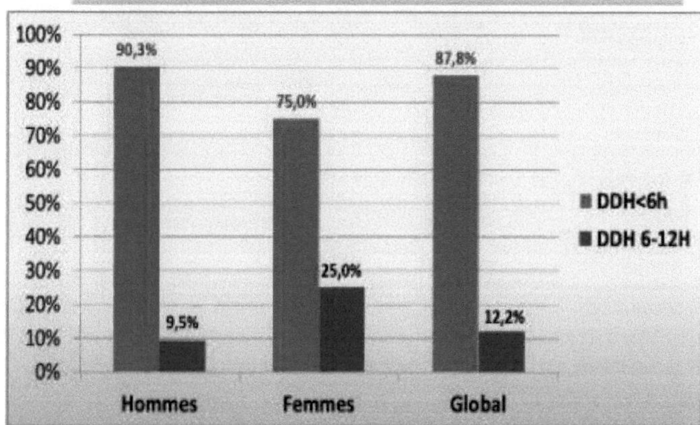

Figure 44: Delays in thrombolysis by sex in relation to the 6th hour

TableXXVII: Time to thrombolysis for patients by age group

Deadlines/Class Age		< 2H	[2H-6H]	< H6	>6H
SERIES	295	94 (31,9%)	157 (53,2%)	251(85,1%)	44 (14,9%)
< 45 YEARS	39	16	18	35	5
[45-54 years]	66	23	31	54	12
[55-64 years]	102	30	56	87	16
[65-75] years	88	25	52	76	11

TableXXVIII: Time and timing of thrombolysis

Thrombolysis		< H6	>H6
DIURNE	216	186 (86,1%)	30 (13,9%)
NOCTURNE	79	67 (84,8%)	12 (15,2%)
TOTAL	295	253 (85,8%)	42 (14,2%)

The [pain-emergency] time was significantly longer (P=0.001) in deceased patients (327 min ± 225 min) than in surviving patients (214 min ± 135 min).

The [pain-emergency] time was significantly shorter ($p < 10^{-3}$) in patients who had undergone successful thrombolysis (192 min ± 118 min) than in patients who had failed thrombolysis (307 min ± 179 min).

c. Different average times for STEMI patients undergoing thrombolysis :

Différents délais moyens des patients thrombolysés

Délais moyens	Minutes	Heures
Global	222	3:42
Consultation directe	137	2:17
Evacuation	270	4:30
Cabinet libéral	327	5:27
Homme	211	3:31
Femme	278	4:38
Diabétique	250	4:10
Non diabétique	211	3:31
Mostaganem	168	2:48
Evacué	261	4:21
Succès	192	3:12
Echec	307	5:07
< 60 ans	205	3:25
> 60 ans	241	4:01
Décédé	327	5:27
Non	214	3:34

Délai d'arrivé tardif des sujets âgés > 60 ans / à ceux < 60 ans : 36min de +

Délai d'arrivée tardif des femmes / aux hommes : 1H07min de +.

Délai d'arrivée tardif des diabétiques / non diabétiques : 39min de +.

Délai d'arrivée tardif des évacués / consultation directe: 2H13min de +

Délai tardif des consultations libérales / directes: 3H10min de +

Figure 45: Time to thrombolysis for STEMI patients in the study

5. Coronary risk factors:

a. Distribution of FDRCs by gender :

Data on coronary risk factors and main medical history were provided for the entire study population. The predominant factors were smoking (64.4%), dyslipidaemia (37.3%), hypertension (37%), diabetes (26.8%) and obesity (10.8%).

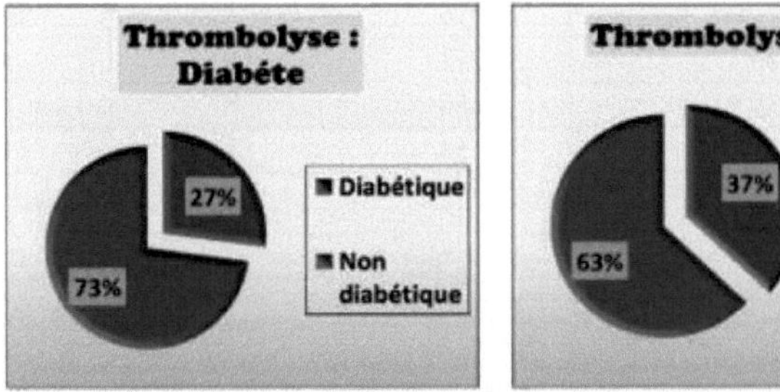

Figure 46: Rate of diabetic and hypertensive patients in the population

Morphological data (weight/height) showed that 45.1% of subjects were overweight (BMI > 25 to 30) and 10.8% obese (BMI > 30).

Compared with men, women were more frequently hypertensive (72.92% versus 29.55%; p=0.02) and more frequently diabetic (62.5% versus 20.24%; p=0.01).

The distribution of coronary risk factors (DRF), established on the basis of anamnestic data, enables the following elements to be selected:

> Age: 76% of male patients were over 50 on the day of the infarction and 75% of female patients were over 60 on the day of the infarction.

> Heredity: Coronary heredity was well established in 35 men and 2 women.

> Hypertension: Known hypertension was found in 30% of male patients and 73% of female patients.

> Diabetes: 20% of male patients and 62.5% of female patients were known to be diabetic on the day of the infarction.

> Smoking: 76.5% of male patients were regular smokers who had not yet given up

smoking on the day of the infarction.

> Combined hypertension and diabetes: 15% of the total number of patients, i.e. 22 men (9%) and 23 women (48%), had both hypertension and diabetes.

> Overweight: 54.65% of men and 33.33% of women were overweight (BMI >25). The mean BMI for diabetics was 26.83 ± 3.24 and for non-diabetics 25.48 ± 3.37 (p=0.002).

> Lipids: dyslipidaemia was identified in 95 men (38.5%) and 15 women (31.25%).

b. Number of FDRCs per patient :

> Only 2 patients (0.7%) had no risk factors on the day of the infarction.

> 64 patients had a single risk factor, i.e. 21.7% of the population.

> 137 patients (46.4%) had 2 risk factors.

> 92 patients, or 31.2% of the population studied, had 3 or more risk factors.

Table XXIX: Number of risk factors per patient and sex

FDRCV	N	%	H	%	F	%
0	2	0.7	0	0	2	4.17
1	64	21.7	53	21.46	11	22.9
2	137	46.4	124	50.2	13	27.1
3 and more	92	31.2	70	28.34	22	45.8

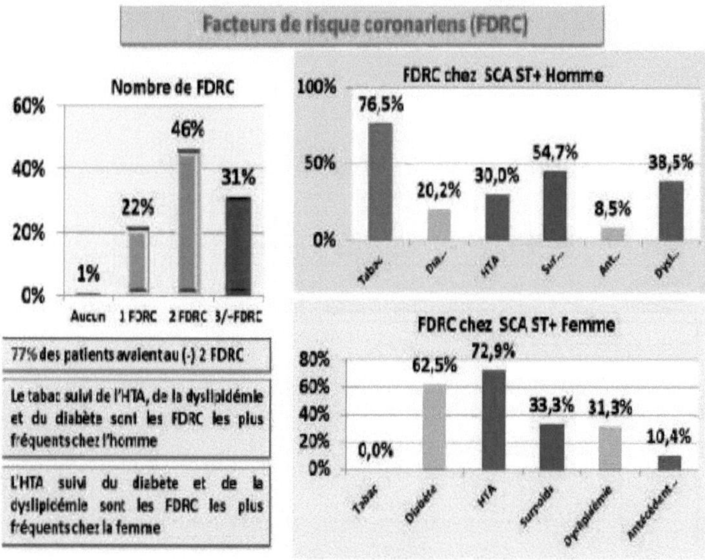

Figure 47: Gender distribution of CRDF in the study population

c. Distribution of CRDF by age :

> The 117 patients aged under 65 were distributed as follows: - 10 had no significant risk factors (8.5%)

- 38 had a single risk factor (32.5%)
- 69 of them had more than 2 coronary risk factors (59%)

> The 32 patients aged 65 and over break down as follows:

- 2 had no known CRDF on the day of the infarction (6%)
- 7 had a single coronary risk factor (22%)

- 23 had more than 2 risk factors (72%).

Table XXIX: Breakdown of the various FDRCs by gender

Gender	Men (%)		Female (%)		Total (%)	
Age	188	76%	36	75%	224	76%
Diabetes	50	20%	30	62.5%	80	27%
HTA	74	30%	35	73%	109	37%
Overweight	135	55%	16	33%	151	51%
Obesity	18	7.3%	5	10.4%	23	7.8%
Tobacco	189	76,5%	0	0	189	64%
Coronary artery disease	21	8,5%	5	0,4%	26	8,8%
Dyslipidemia	95	38.5%	15	31.25%	110	37,3%
Coronary heredity	35	14%	2	4.2%	37	12.5%

5. Initial clinical features :

> . Chest pain :

Pain, the main symptom, is present in 100% of our patients. It is typical in some patients and atypical in others.

> . Hemodynamic status :

The Killip classification is used to assess the haemodynamic status of patients. Killip's (grade I) indicates a stable haemodynamic state, (grade II) reflects latent ventricular failure, while (grades III and IV) are the clinical reflection of a precarious or even threatening haemodynamic situation.

Thus, 247 of our patients (83.73%) were treated with Killip I, 26 (8.8%) with Killip II and 22 patients (7.46%) with Killip III or IV for extensive anterior necrosis.

Table XXXI: Breakdown of KILLIP grades of patients by sex

KILLIP grades /Sex	I		II		[II-IV		P
Male	211	(85,4%)	22	(8,9%)	14	(5,7%)	
Female	36	(75%)	4	(8,3%)	8	(16,7%)	0,029
Total	247	(83,7%)	26	(8,8%)	22	(7,5%)	

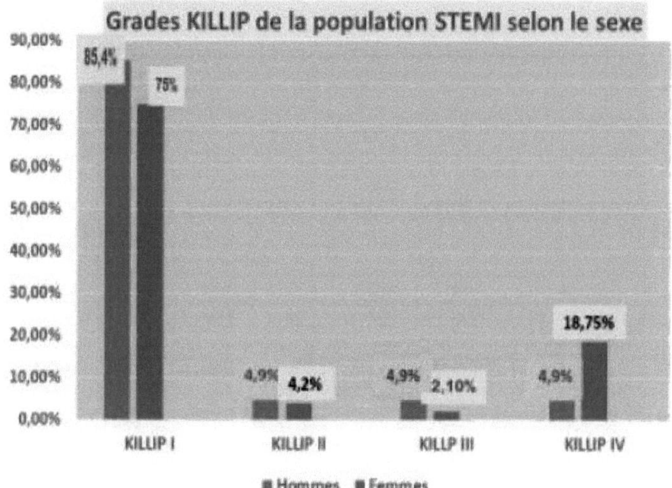

Figure 48: Breakdown of KILLIP grades in the population by gender

- In terms of age groups, of the 22 Killip III or IV patients, 8 are under 60 and 14 are 60 or over.
- Similarly, of the 247 Killip I patients, 129 are under 60 and 118 are 60 or over.
- A significant difference in relation to age was observed when comparing the 3 Killip subgroups I, II and (III, IV): patients aged 60 or over had a worse baseline KILLIP score (p= 0.002).

Table XXXII: Distribution of KILLIP grades of patients by age group

KILLIP/Tranches d'Age	N	I	II	III-IV
SERIES	295	83,7%	8,8%	7,5%
45 YEARS OLD	39	87,2%	7,7%	5,1%
[45-54 years]	66	90,9%	6,1%	3,0%
[55-64 years]	102	82,4%	9,8%	7,8%
[65-75] years	88	78,4%	10,2%	11,4%

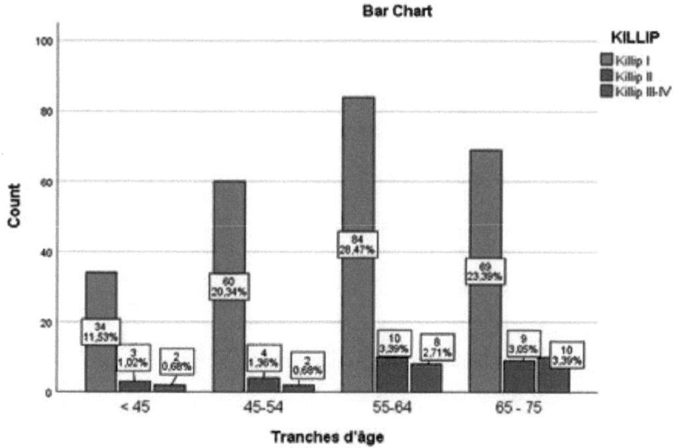

Figure 49: Breakdown of KILLIP grade by age group

> Diabetes: out of 22 Killip III or IV patients, 11 have diabetes and 11 do not.

> When comparing the Killip I, II and (III, IV) subgroups, the Killip grade is significantly worse in diabetics than in non-diabetics (p=0.0001) because of the 11 diabetic patients who have a KILLIP (III, IV), 9 patients are over 60 and only 2 are under 60.

Table XXXIII: Distribution of KILLIP grades of patients and diabetes

Grade KILLIP		I		II		III-IV		P
Workforce Total	295	247		26		22		
DIABETICS	80	64	80%	5	6%	11	14%	
NO Diabetics	215	183	85%	21	10%	11	5%	0.033

Table XXXIV: Distribution of patients' KILLIP grades by anteroinferior territory

Territory	Killip I		Kilip II		Killip III-IV	
Previous 171	147	86%	13	7,6%	11	6,4%
Lower 89	77	86,5%	7	8%	5	5,6%
Other 35	23	66%	6	17%	6	17%
TOTAL	247	84%	26	8,8%	22	7,5%

c.Incidents on admission :

Table XXXV: Incidents in STEMI patients on admission

	ACR	Shock	BAV	TV	OAP	ACFA
Men	2	5	5	3	4	2
Woman	0	3	0	0	0	1
Total	2	8	5	3	4	3

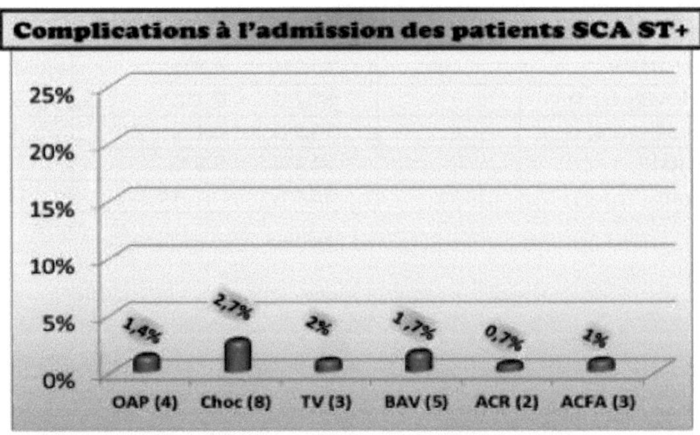

Figure 51: Incidents on admission of STEMI patients

6. Biological parameters on admission :

a. General information :

- hyperleukocytosis (white blood cells greater than 12,000 cells/mm3) was observed in 78 patients, i.e. 26.44% of the population
- low haemoglobin (<14 g/100) in 48 patients (16.3%) with mild thrombocytopenia in five patients (130,000/mm3)

- 80 patients in our series were known diabetics on the day of the infarction. In addition, hyperglycaemia was diagnosed in 8 other patients. In other words, on the day of the infarction, 88 patients had hyperglycaemia, i.e. almost 30% of the population.

b. Myocardial enzymology :

- US troponins were performed in 101 patients, normal in 53 subjects (52.5% of the population) and pathological in 48 subjects (47.5% of the population).

c. Blood crase :

- TP and TCK are normal in 100% of the population.

7. Topography of STEMI :

Anterior necroses predominate in our study, accounting for 59.97% of cases, with anterosepto-apical necrosis being the most frequent (76 cases out of a total of 171 anterior infarcts, i.e. 44.44%). Anterosepto-apical infarction accounted for more than a quarter of necroses in the sample, all topographies combined. In our series, posterior necroses accounted for 41.67% of infarct cases.

Infarct territories

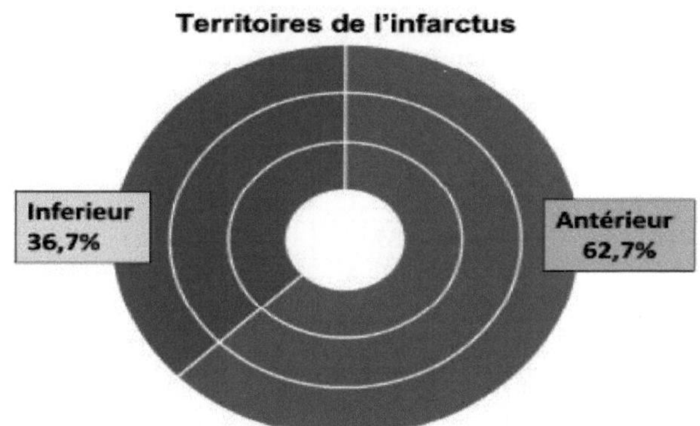

Figure 51: Different anterior and posterior territories of STEMI patients

Table XXXVI: Different anterior and posterior territories in STEMI patients

Territory / Gender		Previous		Lower		P
Male	247	161	65%	86	35%	0,136
Female	48	24	50%	24	50%	
Total	295	171	63%	89	37%	

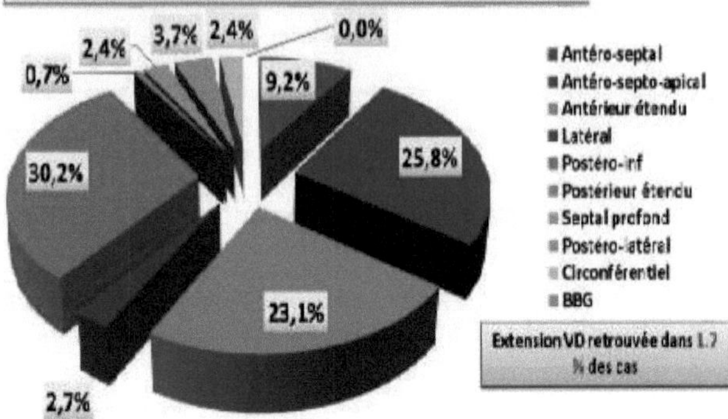

Figure 52: Distribution of STEMI patients by territory

Table XXXVII: Distribution of STEMI patients byTerritory and sex

TERRITORY	N	%	H	%	F	%
ANTEROSEPTAL	27	09,2	21	8.5	6	12.5
ANTEROSEPTOAPICAL	76	25,8	69	27.94	7	14.6
EXTENDED FRONT	68	23,1	59	23.9	9	18.75
PREVIOUS	171	59.97	149	60.3	22	45.83
LOWER	89	30,2	69	27.94	20	41.67
LATERAL	08	2,7	7	2.8	1	2.08
EXTENDED POSTERIOR	02	0,7	2	0.81	0	0
SEPTAL DEEP	07	2,4	7	2.8	0	0
POSTEROLATERAL	11	3,7	8	3.24	3	6.25
CIRCUMFERENTIAL	07	2,4	5	2.02	2	4.17

8. Summary of pre-thrombolysis data :

Table XXXVIII: Summary of pre-thrombolysis data for STEMI patients

Class Age (years)			< 50	50 - 59	60 - 69	70 -75
SERIES		295	54	84	106	41
Men		247	56	80	81	30
Women		48	8	4	25	11
Residence address	Achaacha	6	1	2	2	1
	Ain Nouissy	13	1	3	7	2
	Ain Tadles	24	1	8	11	4
	Bouguirat	29	4	9	11	5
	Kheireddine	15	4	2	5	4
	Mamache	27	4	8	12	3
	Mesra	41	12	13	11	5
	Mostaganem	118	31	32	40	15
	Passenger	09	4	3	2	0
	Sidi Ali	07	0	1	4	2
	Sidi Lakhdar	06	2	3	1	0

Admission procedure	UMC	128	35	38	41	14
	Evacuation	117	25	30	42	20
	Orientation	50	4	16	23	7
Arrival timetable	8am-8pm	216	46	57	82	31
	8pm-8am	79	18	27	24	10
Risk Factors	Diabetes	80	7	22	31	20
	HTA	50	12	22	50	24
	hypertension+diabetes	20	3	8	22	12
	Overweight+Obesity	95	36	51	60	20
	Dyslipidemia	64	24	32	41	13
	Tobacco	64	52	62	59	16
Number of Risk Factors	No	12	1	0	1	0
	1 factor	45	47	9	6	2
	2 factors	49	12	52	57	16
	3 factors and more	43	4	23	42	23
History of angina		26	6	4	12	4
KILLIP III-IV		22	3	5	7	7
Clinical and Electrical Success		220	49	66	82	23
Topography of STEMI	Previous MI	171	30	54	58	19
	Posterior MI	89	16	23	33	17
Echocardiography	Average FEG	49,6	52,58 ± 9.73 %		46,63 ± 10,39 %	
	EF < 40%	36	9	5	13	9

9 Therapeutics :

a. Thrombolysis :

Tenecteplase group (Metalysis): 202 patients in our population received Tenecteplase (68.5%). The mean age in this sub-population was 57.21 ± 10 years.

- **Alteplase group (Actilyse):** 93 patients (21.5%) received Alteplase. Their mean age was 59.25 ± 9 years and they were not significantly older (p= 0.123).

Table XXXIX: Breakdown of Fibrinolytics used / SEX

Gender	Tenecteplase		Ateplase		P
Men	170	68,8%	77	31,2%	0,087
Woman	32	66,7%	16	33,3%	
Total	202	68,5%	93	31,5%	

Table XL: Breakdown of fibrinolytics used by age group

Fibrinolytic	< 45	45-54	55-64	65 - 75	Total	%	P
Tenecteplase	32	46	64	60	202	68.5%	0,003
Alteplase	7	20	38	28	93	31.5%	
Total	39	66	102	88	295	100%	

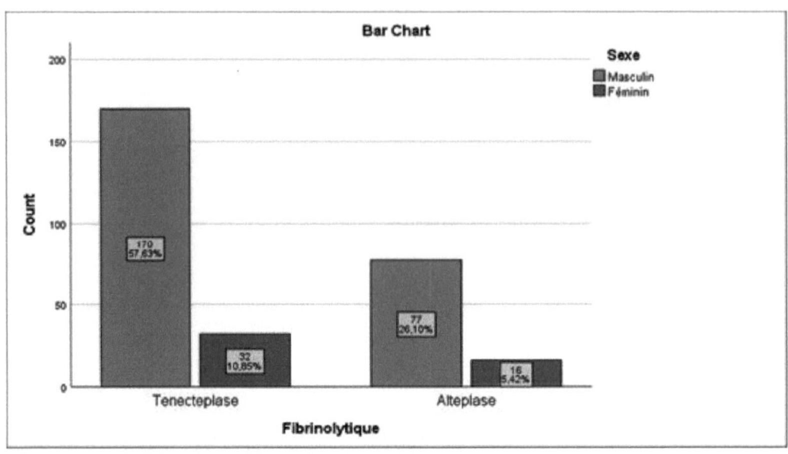

Figure 53: Breakdown of thrombolytics used by sex

b. Associated therapies :

> Enoxaparin: in accordance with the protocol, was administered to all patients.

> Aspirin: in accordance with the protocol, was also administered to all patients either by injection at a dose of 250 mg IVL or 300 mg per os.

> Clopidogrel: in accordance with the protocol, was also administered to all patients at a dose of 300 mg orally.

> Beta-adrenergic blocking agents: 237 patients (80.3%) received beta-adrenergic blocking agents for all myocardial infarction topographies combined. 58 patients had a contraindication to this treatment for the following reasons:

- left ventricular failure: 21 cases
- Two cases of sino-auricular block
- atrioventricular block on extensive anterior necrosis in 3 patients
- cardiogenic shock: 12 cases

> Converting enzyme inhibitors (CEIs): were prescribed in 236 patients (80%).

- 85 times out of a total of 91 previous MI (97.4%)
- 27 times out of a total of 52 posterior necroses (52%)
- 2 times out of 6 for another location.

It was not possible to prescribe them in a number of cases:

- 21 times for blood pressure instability
- 12 times for cardiogenic shock
- 2 times for renal failure
- 4 times for right ventricular extension

> Nitrate derivatives: were used in 239 patients from the start of thrombolytic treatment. 56 patients had contraindications to their use: 12 cardiogenic shocks, 4 right ventricular extensions, 21 IVGs and 15 arterial hypotensions.

Amiodarone: prescribed 3 times for ventricular tachycardia

> Atropine: injected 4 times for vagal discomfort

> Furosemide injection: has been given 8 times to control left ventricular failure or acute lung redema.

> Inotropes positive: 23 times for cardiogenic shock OR decompensated abortion.

> Statins: prescribed to all patients.

Table XLI: Therapeutics associated with in-hospital thrombolysis

PRODUCT	WORKFORCE	%
Beta blockers	237	80,3%
ASPIRIN	295	100%
LMWH	295	100%
IEC	236	80 %
STATINES	275	100%
FUROSEMIDE	8	2,7%
AMIODARONE	3	1%
INOTROPES +	35	11,86%

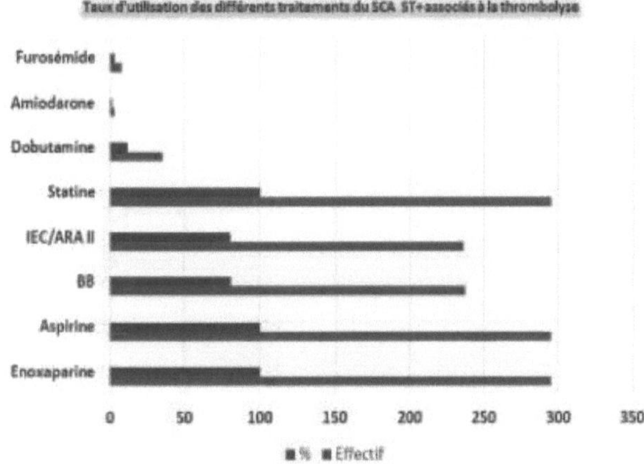

Figure 54: Treatment associated with thrombolysis

10. Short-term trends :

a. Clinical plan :

It concerns developments during the first 48 hours.

Progression of pain: 250 patients out of a total of 295 patients examined (at the end of 48 hours in the ICU) experienced a significant reduction in their pain symptoms before the 3rd hour of thrombolysis.

- Pain remained unchanged in 25 patients (up to 24 hours after thrombolysis).
- Progression of haemodynamic status: At 48 hours, out of a total of 295 patients, 22 remained in a precarious haemodynamic state (Killip III or IV).

Table XLII: Haemodynamic status at 48 hours in STEMI patients

	STATI	UT HEMODYNAM	IQUE A 48 H	
BP mm Hg	PAS<100 admission	PAS<100 (H24)	PAS<100 (H48)	KILLIP > II
WORKFORCE	34	11	10	22
%	11,5%	3,7%	3,4%	7,46%

b. Ultrasound parameters :

b1. Overall results :

> Ultrasound was performed in the cardiology department in 275 patients, usually 48 hours after thrombolysis, and was not performed in 20 patients who died early.

> The mean EF, determined by Simpson's method, is 49.35 ± 10.3% and varies between 20 and 68%. An EF < 40% defines left ventricular dysfunction.

> An ejection fraction < 40% was observed in 36 patients, i.e. 12.2% of the population studied.

> The ejection fraction was also lower in the case of anterior necrosis compared with posterior necrosis (49 ± 9 compared with 52 ± 10). The difference was significant (p=0.002).

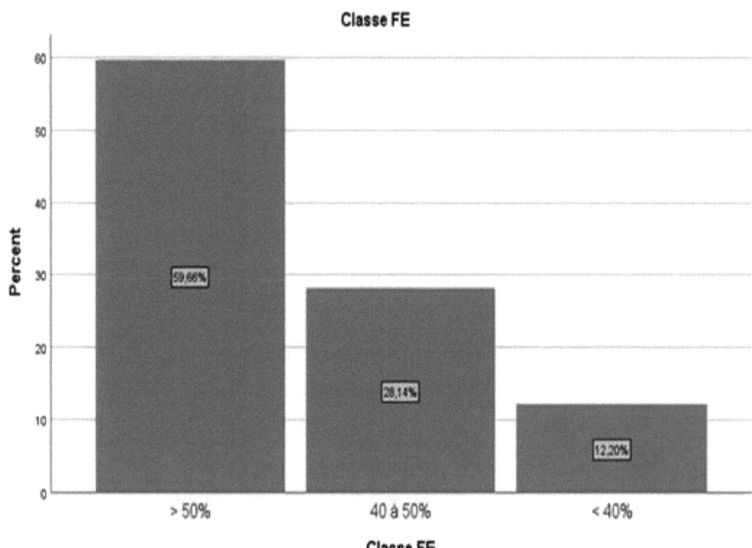

Figure 55: Ejection fraction of STEMI patients

b2. Results by patient category :

> The mean EF in men was 50.1 ± 9.7%; in women it was 45.62 ± 12.5%. The difference was not significant.

> The mean EF was 50.28 ± 10% in non-diabetic patients and 46.85 ± 10.85% in diabetic patients. Significant difference (p=0.011).

> In patients under 65, the mean EF was 51.13 ± 10% compared with 46 ±10 in patients aged 65 to 75. The difference was not significant for this parameter or for LV dysfunction.

> Similarly, for diabetes, there was a significant difference in the number of EFs < 40% (p<0.01).

It should be noted that there was no significant difference in LV dysfunction in relation to age, whether the patients were aged under 65 (20 patients) or over 65 (16 patients).

Table XLIII: Average age of STEMI patients and their LVEF

FE class	Average age	N	%	Standard deviation
> 50%	55,94	176	60%	9,917
40 à 50%	61,31	83	28%	9,834
< 40%	59,22	36	12%	12,681
>40	57,66	259	88%	10,185

Total	57,85	295	100%	10,509

Table XLIV: LVEF of STEMI patients by sex

FE/Sex class	> 50%	40 à 50%	< 40%	Total
Male	152	68	27	247
%	61,5%	27,5%	10,9%	100,0%
Female	24	15	9	48
%	50,0%	31,3%	18,8%	100,0%
Total	176	83	36	295
%	60%	28%	12%	100,0%

Table XLV: LVEF of STEMI patients by age group

FE classes / age (years)	< 45	45-54	55-64	65-75	TOTAL
> 50%	25	52	62	37	176
%	14,2%	29,5%	35,2%	21,0%	100,0%
40 à 50%	8	8	31	36	83
%	9,6%	9,6%	37,3%	43,4%	100,0%
< 40%	6	6	9	15	36
%	16,7%	16,7%	25,0%	41,7%	100,0%
Total	39	66	102	88	295
%	13,2%	22,4%	34,6%	29,8%	100,0%

Table XLVI: LVEF classes 40% of STEMI patients by age group 65 years

FE class	EF < 40%	EF > 40%	P
PATIENTS < 65 years	20	172	
PATIENTS > 65 years	16	87	0 ,012
TOTAL	36	259	

Table XLVII: LVEF 40% of STEMI patients according to anteroinferior STEMI topography

FE class	EF > 40%	EF < 40%	TOTAL
Previous	153	18	171
Lower	82	7	89
Other	24	11	35
TOTAL	259	36	295
P	0,001		

c.TIMI score :

-129 patients, i.e. 43.73% of the total, were classified as **TIMI 0, 1 or 2**; 122 of them (94.6%) were under 65.

-108 patients were classified as **TIMI 3, 4 or 5**, i.e. 36.6% of the total; 52 of them (48%) were under 65.

-58 patients **had TIMI>6** scores, i.e. 19.66% of the total; 33 of them (57%) were under 65 and 25 (43%) were over 65.

-Just over half the patients (164 or 55.6%) had a TIMI score of 1 to 3.

Table XLVIII: STEMI patients according to TIMI risk score

TIMI	0	1	2	3	4	5	6	7	> 8
Total	15	52	62	50	27	31	24	15	19
%	5	17,6	21	17	9	10,5	8	5	6,4

Brackets	43,73 %	36,6%	19,66 %

Table XLIX: STEMI patients according to TIMI risk score by age group

Age groups	SCORES TIMI								
	0	1	2	3	4	5	6	7	> 8
< 45 YEARS N=39	2	12	8	6	3	0	5	3	0
[45-54 years] N=66	5	21	19	10	5	2	3	0	1
[55-64 years] N=102	8	17	30	19	2	5	9	4	8
[65-75 years] N=88	0	2	4	15	17	24	8	8	10
TOTAL N=295	15	52	61	50	27	31	25	15	19

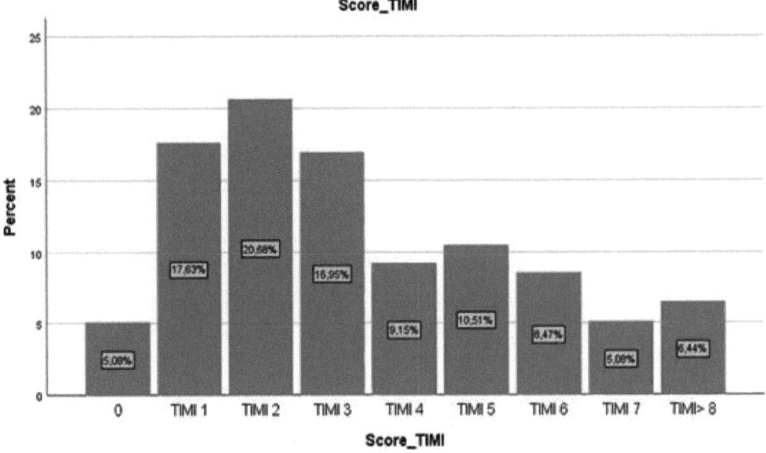

Figure 56: Distribution of TIMI Score in thrombolysed STEMI patients in the study

- Length of stay in ICU: This has a definite prognostic value. It is the initial evolution during the first 48 hours that determines the length of hospitalisation in the ICU.
- 220 people had a favourable outcome, without complications, and were discharged from the ICU after 48 hours. This represents 74.6% of the total population.
- 10 patients (3.4%) stayed in the ICU for more than 48 hours because of a refractory abortion under haemodynamic support.
- 19 patients (6.4%) had a shortened hospital stay of less than 24 hours due to death.
- 45 patients were evacuated or referred for salvage angioplasty after failed thrombolysis.

Table L: Progression of STEMI patients in the ICU at 48 hours

Evolution at 48H	Deaths <48H	Transferred to cardiology	Oriented for angioplasty	Kept in USIC
WORKFORCE	20	220	45	10
%	6,78%	74,6%	15,25%	3,4%

d. Success of thrombolysis and clinical efficacy :

It is defined by the onset of a reperfusion syndrome which combines the disappearance of pain and the regression of ST+.

Pain regression: was found in 250 of our patients

Normalisation of the ST+ segment at 180 minutes was observed in 220 patients.

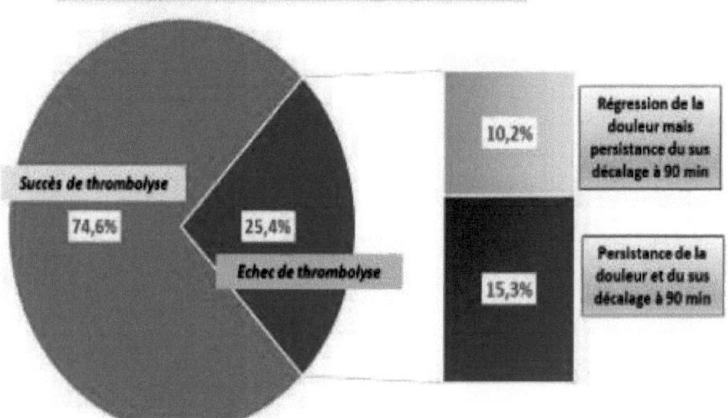

Figure 57: Overall outcome of thrombolysed patients in the study

e. Reperfusion syndrome and age groups :

The disappearance of chest pain and the normalisation of the ST were more rapid in patients under 60 than in those over 60.

In summary, patients who had reperfused were significantly younger than those who had not, (57.26 ± 14.14 years) compared with (59.59 ± 11.42 years) with a (P=0.098).

Of the 220 patients who reperfused (with clinical and electrical success), 55 were diabetic (68.75% of a total of 80 diabetic patients) whereas 165 patients who reperfused were non-diabetic (76.74% of a total of 215 non-diabetic patients) with a P of (0.002).

Similarly, in the total sample of 295 patients, in the sub-group of 220 patients who reperfused, there were 55 diabetic patients (i.e. 18.6%) and 165 non-diabetic patients (i.e. 55.9%) with a P of (0.003).

Mean thrombolysis times were significantly shorter in the reperfused subgroup compared with the non-reperfused subgroup, 202 min ± 7 vs 318 min ± 20 (P<0.05).

Table LI: Clinical reperfusion rates in thrombolysed STEMI patients

reperfusion		YES N=220	NO N=75	P
Average age		**57.26 ± 10.14 years**	**59.59 ± 11.42 years**	0,098
< 60 years	148	**115**	**33**	
> 60 years	147	**105**	**42**	
Diabetics	80	**55**	**25**	
Non-diabetics	215	**165**	**50**	0,003
Angina	26	**8**	**18**	-
Previous Idm	171	**130**	**41**	-
Lower Idm	89	**77**	**12**	-
Time to thrombolysis		**202 min ± 7 min**	**318 min ± 20 min**	<0,05
< 2H	94	**79**	**15**	
] 2H-6H]	157	**125**	**32**	
>6H	44	**16**	**28**	0,017

f. Effectiveness and time to thrombolysis :

The evolution of ST elevation according to the time taken for thrombolysis shows that "early" thrombolysis before H6 is very significantly associated (p=0.0001) with better ST+ dynamics at 180 min.

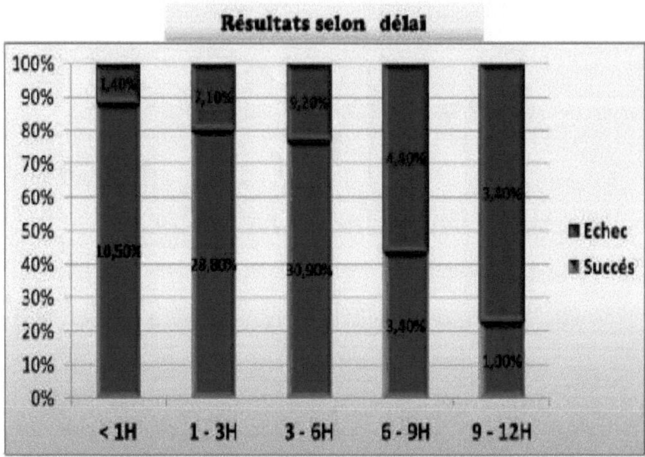

Figure 58: Breakdown of results according to thrombolysis delay

Table LII: ST+ segment dynamics at 90min and time to thrombolysis

Time to thrombolysis		Resolution ST+ N=220 (74,6%)	Persistent ST+ disease 90 min N=63 (21.35%)
<2H	N=94	79	15
[2H-6H]	N=157	125	32
>6H	N=44	16	28
Thrombolysis< H6	253	205	48
Thrombolysis> H6	42	15	27

g. Successful thrombolysis and left ventricular dysfunction :

Complete reperfusion was achieved mainly when patients were thrombolysed before H6 (85.71%).

It should be noted that 18 patients continued to have LV dysfunction despite successful thrombolysis and therefore constitute a sub-group of patients (success + EF<40%).

In this group of 18 patients (success+ FE<40%), there were 13 diabetics compared with 5 non-diabetics. The difference is significant.

Of these 18 patients, 8 underwent fibrinolysis after H6 compared with 10 before H6, a highly significant difference (p=0.0008). For TIMI3 patients, thrombolysis after H6 appears to have a poor prognosis.

Table LIV: LVEF 40% and diabetes

	EF<40%. N=36		EF>40% N=259		P
DIABETICS	11	13.75%	69	86.25%	
NON-DIABETICS 215	25	11.63%	190	88.4%	0,621

Table LIV: LVEF 40% and time to thrombolysis

	EF<40% N=36	EF>40% N=259	P
Thrombolysis < H6 253	25	228	
Thrombolysis >H6 42	11	31	0,003

i. Clinical reperfusion according to thrombolysis schedule :

There was no significant difference between patients undergoing thrombolysis at night or during the day. Similarly, there was no statistically significant difference in coronary patency between patients arriving on a working day and those arriving on a bank holiday.

Table LV: Total thrombolysis success and time of admission for STEMI patients

REPERFUSION	Daytime thrombolysis 216		Nocturnal thrombolysis 79	
Success Total N=220	165	55,9%	55	18,64%
Clinical success N=250	186	63%	64	21,7%

h. Electrocardiographic :

220 patients with a favourable ST+ progression at the 90th minute, while 30 patients, i.e. 10% of the initial group, still had a significant ST+ elevation at 3 hours after thrombolysis despite the disappearance of pain. These were the patients for whom nitrate derivatives were maintained for up to 48 hours. Analysis of the factors likely to influence ST+ kinetics at 180 minutes incriminates the delay in thrombolysis, age and diabetes.

Table LVI: Possible determinants of ST+ normalisation at 90 min

		Electrical success N=220	Electrical failure N=63	P
THROMBOLYSIS < H6	253	205	48	
THROMBOLYSIS > H6	42	15	27	0,0001
AGE <65ANS	192	147	45	
AGE >65ANS	103	73	30	0,285
DIABETE	80	54	26	
WITHOUT DIABETES	215	166	49	0,089
PREVIOUS IDM	171	130	41	0,0001
IDM POST	89	77	12	

Table LVII: Clinical and electrical success rates of thrombolysis

Success		Clinique Seul	%	Clinical + electrical	%	P value
Yes		250	84,7	220	74,6	
No		45	15,3	75	25,4	0,0001
Men		213	86.2	189	76.5	
Woman		37	77.1	31	64.6	0,004

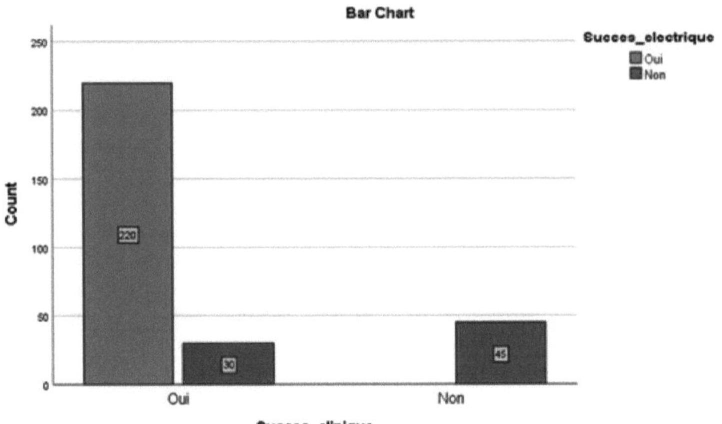

Figure 60: Breakdown of clinical and electrical success rates

Table LVIII: Clinical and electrical thrombolysis success rates by age group

Success / Age group	< 50	50 - 60	60 - 70	70 - 75	Total
Total success	49	66	82	23	220
%	16.6	22.4	27.8	7.8	74.6%
Clinical success alone	8	7	6	9	30
%	2.7	2.4	2	3	10.15%
Total failure	7	11	18	9	45
%	2.4	3.7	6.1	3	15.25%

Table LIX: Summary of STEMI patient outcomes at 48 hours

	ADMI	[SSION- THRC		1MBOLYSE		A 48	HOURS			
Age groups (years)	N	< 45	45 - 54	55 - 64	65 - 75	N	< 45	45 - 54	55 - 64	65 - 75
WORKFORCE	295	67	50	29	3	55	8	14	15	18
Men	247	36	59	86	66	46	8	11	12	15
Women	48	3	7	16	22	9	0	3	3	3
Pain	295	67	50	29	3	25	3	5	10	7
Sus ST shift	295	67	50	29	3	55	8	14	15	22
KILLIP III-IV	22	2	2	8	10	6	0	2	2	2
TIMI > 6	61	8	5	22	26	44	6	5	15	18
EF < 40	36	6	6	9	15	18	4	6	2	6

11. Early complications in thrombolysed STEMI patients :

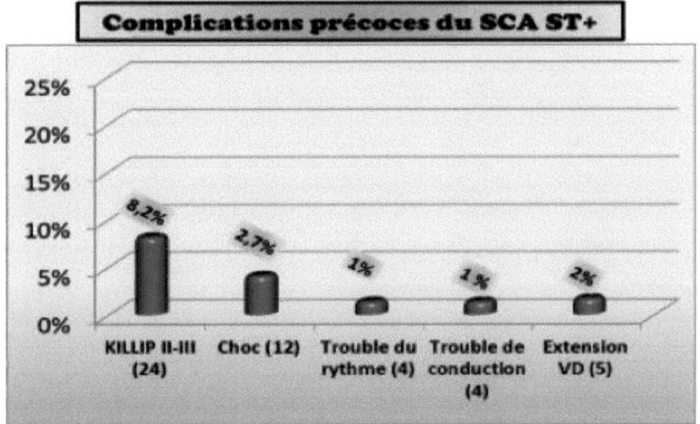

Figure 61: Early complications in thrombolysed STEMI patients

Table LX: Early complications in thrombolysed STEMI patients

	WORKFORCE		MEN N=247	WOMEN N=48
TV	3	1%	**7**	**2**
BAV II-III	5	1.7%	**3**	**2**
ABORTION	21	7.12%	**17**	**4**
CARDIOGENIC SHOCK	12	4%	**9**	**3**
Persistence of ST+ at 180 minutes	75	25.4%	**58**	**17**
Extension VD	4	1.35%	**1**	**3**
KILLIP	3	1%	**3**	**0**

Table LXI: Early complications in STEMI patients according to age

Age groups of STEMI patients	< 45	45-54	55-64	65 - 75	Total
CARDIOGENIC SHOCK	1	0	9	2	12
Shock + VD extension	0	0	0	4	4
ABORTION	3	4	6	8	21
KILLIP	0	2	0	1	3
Persistence of ST+ at 180 minutes	10	14	24	27	75
BAVII-III	1	0	0	4	5
TV	0	0	2	1	3
Total	39	66	102	88	295

Table LXII: Early complications in STEMI patients and diabetes

early complications	Total 123	Diabetes n=41	No diabetes n=82
CARDIOGENIC SHOCK	12	4	8
Shock + VD extension	4	4	0
ABORTION	21	4	17
KILLIP	3	1	2
TV	3	0	3

BAVII-III	5	2	3
Persistence of ST+ at 180 minutes	75	26	49

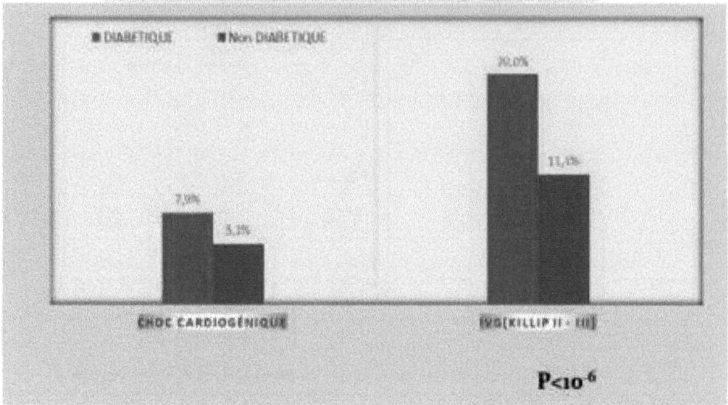

Figure 61: Impact of diabetes on the outcome of STEMI patients

<u>Table LXIII: Early complications by age 65</u>

Early complications of STEMI compared to age 65			
	< 65	65-75	Total
CARDIOGENIC SHOCK	9	3	12
Shock + VD extension	0	4	4
ABORTION	13	8	21
KILLIP	2	1	3
TV	2	1	3
BAVII-III	1	4	5
Persistence of ST+ at 180 minutes	45	30	75

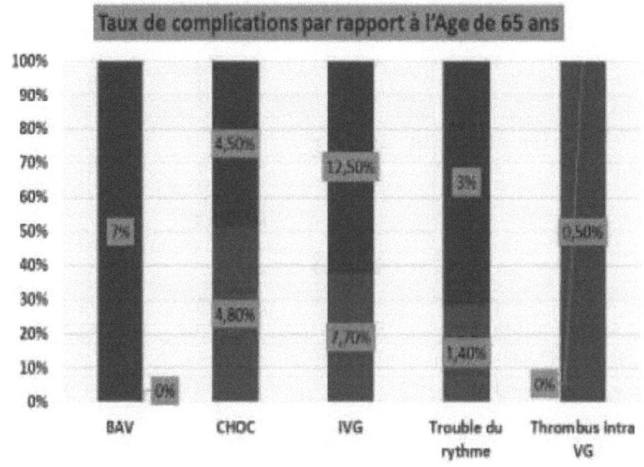

Figure 62 : Complication rates in relation to age 65

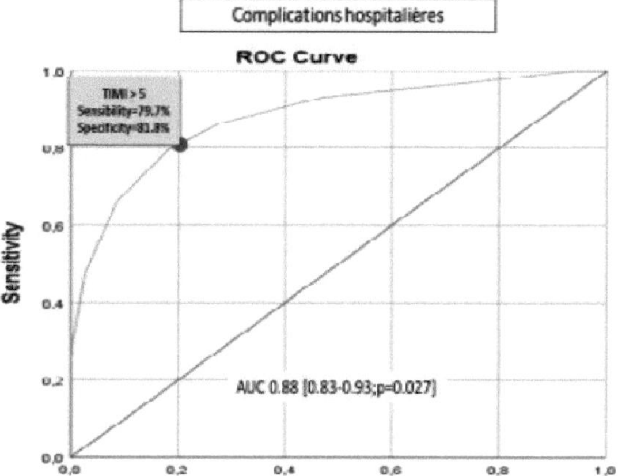

Figure 63: ROC curve for hospital complications

Table LXIV: Univariate and multivariate study of hospital complications

'I para Nation dun rr-tr пи di thrnmbnlyif of the syndrome nprnariín aiju лт4 tux urline" mitìw th¡rurgicalí t de Mostagán oт

"Installation d'un réseau de thrombolyse du syndrome coronarien aigu ST+ aux urgences médico-chirurgicales de Mostaganem"

Complications hospitalières/Paramètres	Analyse univariée OR [IC 95%] p	Analyse multivariée ORa [IC 95%];p
Age : 65-75 ans	2.1 [1.1-3.8]; p=0.017	2.6 [1.0-6.5] ; p= 0.043
Sexe féminin	3.9 [1.5-10.2]; p= 0.003	2.9 [3.9-9.1] ; p= 0.072
Provenance	2.2 [1.2-4.1]; p=0.017	4.1 [1.5-10.7] ;p= 0.004
Délai ≥ 6H	4.9 [1.5-12.7]; p< 10⁻³	0.5 [3.2-1.7] ; p= 0.258
Diabète	2.9 [1.2-7.3]; p=0.017	1.5 [0.2-1.6] ; p= 0.257
HTA	2.7 [1.1-6.9]; p=0.027	0.6 [0.2-1.9] ; p= 0.400
Antécédent ischémique	5.4 [1.9-15.7]; p=0.001	3.4 [0.9-11.1] ; p= 0.078
FC>100 bats/min	14.3 [7.2-28.6]; p< 10^{-3}	8.6 [3.2-22.5] ;p< 10^{-3}
3 FCRC et plus	4.6 [1.8-12.0]; p= 0.001	1.9 [0.5-6.7] ;p=0.317
Echec de thrombolyse	20.9 [10.3-42.5]; p< 10^{-3}	10.0 [4.0-25.0] ;p< 10^{-3}

12. Post-thrombolysis accidents and incidents :

Table LXV: Haemorrhagic complications of thrombolysed spacecraft

	SERIES	%
Minor and minimal bleeding	6	2%
Major bleeding	2	0,7%
TOTAL	8	2,7%

Table LXVI: Bleeding complications and age groups of STEMI patients treated with thrombolysis

Class Age	< 45	45-54	55-64	65 - 75	Total
Epistaxis	1	0	0	0	1
Haematemesis	1	0	0	0	1
Gingivorrhagia	0	1	1	0	2

Haematuria	0	0	1	0	1
Brain haemorrhage	0	1	0	0	1
Minimal haemorrhage on AVF	0	0	1	0	1
Haemorrhagic Syndrome	0	0	1	0	1
TOTAL	2	2	4	0	8

Table LXVII: Post-thrombolysis accidents and incidents

Thrombolysis accidents and incidents

Accidents et incidents à la thrombolyse

N=295	Nombre total d'accidents	Hémorragies Mineures	Hémorragies majeures
N	9	7	2
%	3,05 %	2,37 %	0,68 %

13. Study of in-hospital mortality :

a. One-month mortality rate :

Of a total of 295 patients, 20 died between 0 and 30 days.

Total mortality was therefore 6.78%. At H6 the number of deaths reached 18 patients (i.e. 6.1%), at 24 hours we recorded 1 more death (i.e. 6.4%) and at 48 hours we had another death, giving a total of 20 patients who died and 275 who survived.

Table LXVIII: Mortality rate of STEMI patients treated with thrombolysis

Deceased	N	A H6	From H6 to 24 H	From H24 to D30
SERIES	20	18	1	1
%	6,78%	6,1%	0,33%	0,33%

b. Clinical characteristics of patients who died :

> Age: the mean age of patients who died was 63.5 ± 9.9 years, compared with 57.44 ± 10.4 years for survivors.

> Patients who died were significantly older than those who survived (p = 0.013).

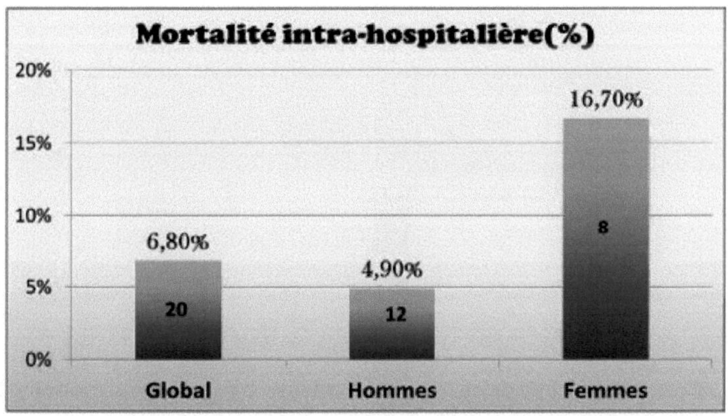

Figure 65: Overall mortality by sex in the STEMI population

- In terms of the age groups of patients who died, 2 were aged under 40, 9 between 55 and 64 and 9 between 65 and 75.
- Gender: 12 men and 8 women died.
- The men who died: 1 aged under 45, 7 aged 55 to 64 and 4 aged 65 to 75.
- The women who died: 1 aged under 45, 2 aged between 55 and 65 and 5 aged between 65 and 75.
- Topography of the myocardial infarction: 8 of the 20 patients who died had an infarction with an extensive anterior topography, 5 patients had an inferior topography and 3 patients had a circumferential topography.
- Diabetes: 10 out of 20 patients who died had diabetes.
- Haemodynamic status: 16 patients (8 men and 8 women) out of 20 who died had a precarious baseline haemodynamic status (KILLIP III or IV).
- Thrombolysis time: 9 out of 20 patients who died were thrombolyzed after H6. The average thrombolysis time for patients who died was 327 min (5H27min) compared with 214 min (3H34min) for survivors.
- At H6, 15 patients had died and after 6ème hours, 5 other patients had died, giving a total of 20 STEMI patients who had been thrombolysed on time and who had died (table 66).
- Mortality and TIMI risk score: 17 patients who died out of a total of 20 had a TIMI score >6; 6 of these were scored (6, 7 or 8) and the remaining 11 patients had a score greater than 8.

Table LXIX: Mortality of thrombolysed STEMI patients by sex and age

Deaths/Gender/Age (years)	< 45	45-54	55-64	65 - 75	TOTAL
Men	1	0	7	4	12
Woman	1	0	2	5	8
Total	2	0	9	9	20

Table LXX: Mortality of thrombolysed STEMI patients and their TIMI Score

Death: TIMI risk score	WORKFORCE
0, 1, 2	1
3, 4,5	2

6, 7, 8	6
> 8	11

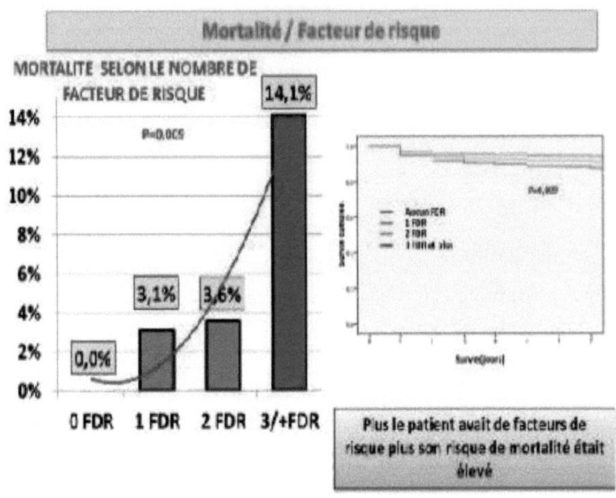

Figure 66 : Mortalité selon le nombre de FDRC

Mortalité Hospitalière

ROC Curve

Diagonal segments are produced by ties.

Figure 67: ROC curve for hospital mortality

Table LXXI: Summary data for patients who died

Summary of deceased patients								
N°	Code	Territory	Age (years)	Deadline	Nb FRCV	Fibrinolytic	Time of death	CAUSE
1	113	Extended front	61	H2	3	Alteplase	3H	SHOCK
2	214	Extended front	63	H10	3	Alteplase	2H	SHOCK
3	223	Extended front	63	H5	2	Tenecteplase	6H	SHOCK
4	257	Extended front	39	H2	1	Tenecteplase	1H	SHOCK
5	292	Extended front	59	H4	5	Alteplase	4H	SHOCK
6	270	Extended front	38	H7	1	Alteplase	3H	SHOCK
7	230	Lateral	66	H7	2	Alteplase	1H	SHOCK
8	249	Anteroseptoapical	59	H3	2	Tenecteplase	2H	SHOCK
9	231	Circonferential	64	H10	5	Alteplase	4H	SHOCK
10	21	Inferolateral	74	H11	2	Alteplase	4H	SHOCK
11	244	Lower	67	H9	3	Tenecteplase	2H	BAV
12	248	Inferolateral	70	H3	3	MET	1H	BAV
13	252	lower	74	H2	3	MET	1H	BAV
14	265	circumferential	69	H1	4	MET	1H	BAV

15	141	extended front	61	H6	4	MET	2H	TSV
16	106	lower	75	H1	2	MET	30 minutes	TV
17	69	extended front	61	H11	3	ACT	6H	OAP
18	132	circumferential	70	H8	3	MET	39H	OAP
19	283	lower	73	H11	3	ACT	6H	OAP
20	138	lower	64	H2	3	MET	12H	Hemorr agie

Table LXXII: Univariate and multivariate study of hospital mortality

Morti liti hfljpiti liH/PlFi rritw	Anular unlwttf OB[K9S%j;p	Analpt 1лиМ|цг1й wn^tifln nell- 0Ri [FC9S%J;p
Age.tfS-TSMt	2.D [П.В-5.1]; p= 0.134	-
Sene iémEnEn	3.9[1.HD.Z]j piD.007	
Source	4,911,3-155]; p*Я7	
	4Í [1.7-124]; p=fl.005	Ц [Ш9.6]; p=M2
Dii bèta	2.9 [1.2-7.3] ¡p= M3	
НТД	2.7 | LI-TJDJ; p= Ü.D32	
Anríiídentlwhémlque	9.511.9-15.4]; p=0.ДО4	
3 FDRC and plui	4.6 [1£Ш]; p=O.OH	
FC>100 bati/mrn	14.115.1-38.9); p <1D-4	
E(h "dethr&mbttlyw	Hl[9J "7.ÖJ;p<№*	-
Cardiogenic shock	271,1[ЗШ1Ш]!p<10	MSJ [35.2-25Д2.4]; p < ID
Rhythm disorder	7.У [l.a-34jl]¡ p =Ü.O17	▪

Table LXIII: Breakdown of deaths over time by dominant cause and thrombolytic agent

Time to death after thrombolysis	0:30	1:00	2:00	3:00	4:00	6:00	12:00	39:00	Total	P
Tenecteplase	1	4	3	0	0	1	1	1	11	0,08
Alteplase	0	1	1	2	3	2	0	0	9	
Etat de Choc	1	2	4	2	0	3	0	1	13	0,000
Shock-free	0	3	0	0	3	0	1	0	7	
Workforce	1	5	4	2	3	3	1	1	20	

> At H24, 6 patients had died in cardiogenic shock following extensive anterior infarction and 2 patients from malignant ventricular fibrillation refractory to resuscitation manoeuvres and unresponsive to amiodarone-based treatment.

> Between H24 and 1 month, there was 1 death, that of a 70-year-old patient who presented with circumferential necrosis complicated by a PAO associated with a state of shock and finally asystole and death after 39 hours.

Table LXXIV: Clinical characteristics of patients who died

Death classes	<H6	H6 to J1	D1 to D30
Workforce	N=18	N=1	N=1
AVERAGE AGE (years)	59,9± 13,3	64	70
FEMALE SEX	3	4	1
PREVIOUS IDM	6	3	0
IDM INFERIOR	4	2	0
TABAC	8	3	0
DIABETE	4	5	1
KILLIP I, II	3	2	0
KILLIP III, IV	9	5	1

PAS<100	6	4	0
HR>100	6	6	1
Thrombo delayyse			
<2H	8	0	0
] 2H-6H]	5	1	0
> 6H	7	6	1
<H6	13	1	0

Table 75

c. Causes of death :

Table LXXV: Causes of death in thrombolysis patients

	Death before H24 N=19	Death between H24 and D30 N=1
BAV-ASYSTOLIA	4	0
TV-FV	2	0
SHOCK	10	0
Massive OAP	2	1
Haemorrhage	1	0

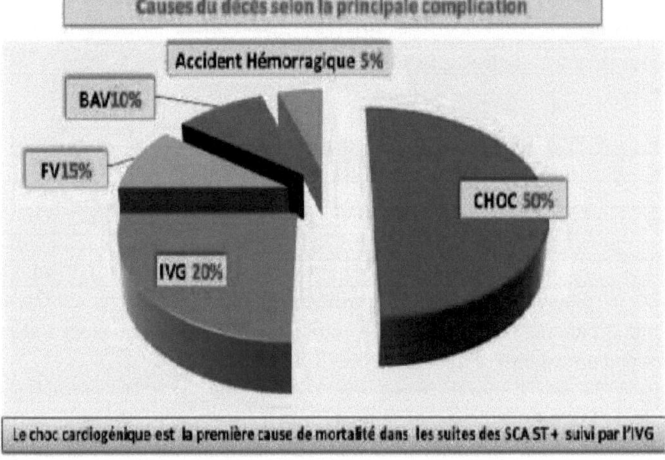

Figure 68: Causes of death in the STEMI population

Table LXXVI: Mortality and time to thrombolysis

Time to thrombolysis	H0-H2	H2-H6	>H6
Deaths	8	5	7
Mortality	2,7%	1,7%	2,4%

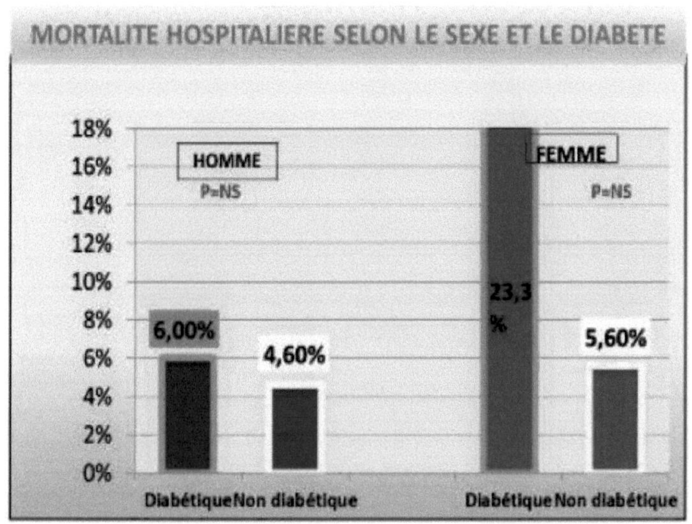

Figure 69: Mortality by diabetes and sex in the STEMI population

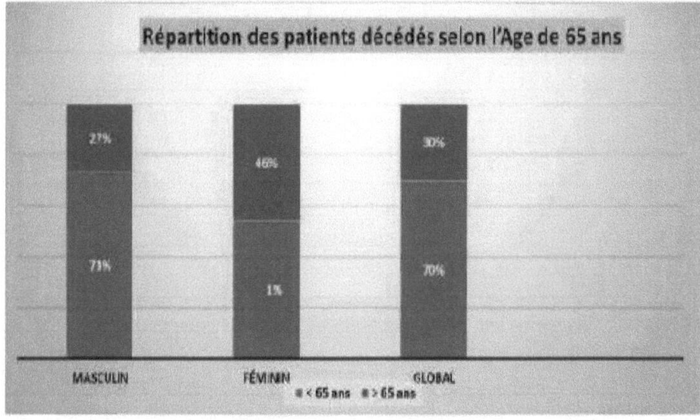

Figure 70: Breakdown of patients who died by age 65

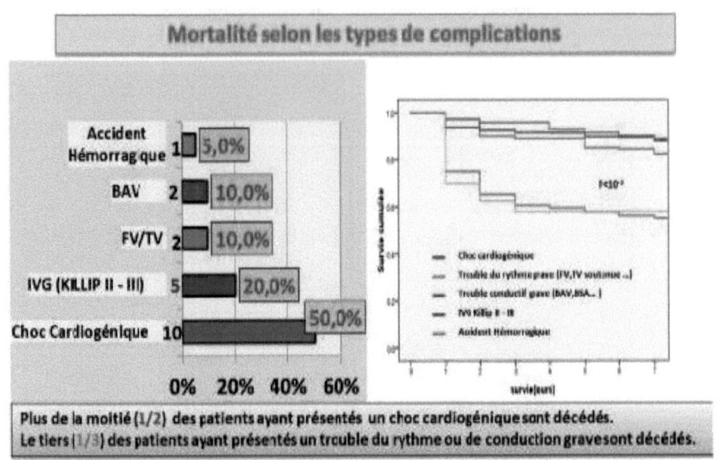

Figure 71: Mortality by type of complication

14 Patient discharge :

Table LXXVII: Discharge mode of thrombolysed STEMI patients by gender

Discharge from the ICU	Success +Transfer to Cardiology	Failure+ Referral for angioplasty	Deceased
H	177	58	12
%	60%	19.7%	4.07%
F	31	9	8
%	10.5%	3%	2.7%
TOTAL	208	67	20
%	70.5%	22.7%	6.78%

Table LXXVIII: Discharge mode of thrombolysed STEMI patients by age group

Exit mode/Class Age	< 45	45-54	55-64	65 - 75	Total	%
Stable in Cardiology	29	52	76	51	208	70.5%
Referred for angioplasty	8	14	17	28	67	22.7%
Deceased	2	0	9	9	20	6.8%
Total	39	66	102	88	295	100

On the discharge prescription, the prescriptions can be detailed as follows:

> Enoxaparin: a curative dose was prescribed for all patients.

> Aspirin: 100 mg per day orally was prescribed to all patients .

> Clopidogrel: 75 mg orally per day was prescribed for all patients.

> Beta-adrenergic blocking agents: were prescribed in 237 patients (80.3%), except in 58 patients who had contraindications to this treatment for the following reasons (21 cases of abortion, 5 cases of AVB, 12 cases of cardiogenic shock).

> Converting enzyme inhibitors (CEIs): were prescribed for 236 patients (80%), except for 39 patients who were contraindicated for the following reasons (21 cases of blood pressure instability, 12 cases of cardiogenic shock, 2 cases of renal failure, 4 cases of right ventricular extension).

> Nitrate derivatives: were prescribed for 239 patients, except for 56 patients with contraindications to their use (12 cardiogenic shocks, 4 right ventricular extensions, 21 IVGs and 15 arterial hypotensions).

> Amiodarone: prescribed 3 times for ventricular tachycardia

> Furosemide per os: has been prescribed 8 times for abortion patients.

-statins: prescribed to all patients.

Table LXXIX: Discharge medical treatment

Discharge medical treatment			
	Medication	**Workforce**	**%**
Discharge medical treatment	**Beta blockers**	**237**	**80,3%**
	IEC	**236**	**80%**
	Nitrates	**239**	**81%**
	STATINS	**275**	**93.2%**
	Aspirin	**275**	**93.2%**
	AMIODARONE	**3**	**1%**
	Anticalciques	**2**	**0.7%**
	FUROSEMIDE	**8**	**2.7%**

14. Overall summary of STEMI patient data :

Table LXXX: Overall summary of STEMI patient data from the study

Class Age (years)			**< 50**	**50 - 59**	**60 - 69**	**70 -75**
Deceased		**20**	**2**	**2**	**10**	**6**
Men		**12**	**1**	**2**	**7**	**2**
Women		**8**	**1**	**0**	**3**	**4**
Residence	**Mostaganem**	**3**	**0**	**0**	**1**	**2**
	Ain Nouissy	**1**	**0**	**0**	**1**	**0**
	Ain Tadles	**4**	**1**	**0**	**3**	**0**
	Bouguirat	**4**	**1**	**0**	**2**	**1**
	Sidi Ali	**1**	**0**	**1**	**0**	**0**
	Mamache	**4**	**0**	**0**	**2**	**2**
	Mesra	**2**	**0**	**0**	**1**	**1**
	Passenger	**1**	**0**	**1**	**0**	**0**
Admission method	**Emergency positions**	**6**	**0**	**1**	**3**	**2**
	Evacuation	**9**	**2**	**1**	**5**	**1**
	Orientation	**5**	**0**	**0**	**2**	**3**
Admission time	**8am-8pm**	**15**	**1**	**2**	**7**	**5**
	8pm-8am	**5**	**1**	**0**	**3**	**1**
Risk Factors	**Diabetes**	**10**	**0**	**1**	**4**	**5**
	HTA	**11**	**0**	**1**	**7**	**3**
	Obesity	**4**	**0**	**1**	**1**	**2**
	Overweight	**7**	**1**	**0**	**4**	**2**
	Dyslipidemia	**14**	**1**	**1**	**8**	**4**
	Tobacco	**11**	**1**	**2**	**7**	**1**
Number of Risk Factors	**No**	**0**	**0**	**0**	**0**	**0**
	1 factor	**2**	**2**	**0**	**0**	**0**

	2 factors	5	0	1	2	2
	3 factors and more	13	0	1	8	4
History of angina		6	1	0	4	1
Killip III-IV		16	2	2	6	6
TIMI > 5		17	2	1	9	5
Topography of STEMI	Previous	9	2	2	5	0
	Posterior	5	0	0	2	3
	Posterolateral	2	0	0	0	2
	Circumferential	3	0	0	2	1
Delay Pain - Thrombolysis		12	1	1	7	3
Echocardiography	EF < 40%	18	2	1	9	6

15. Impact of the installation of a STEMI thrombolysis network in the emergency department of the Mostaganem University Hospital :

The installation of a thrombolysis network in the emergency ICU of Mostaganem has been very judicious and of great therapeutic and prognostic interest for the population of the region, since from year to year the recruitment of patients is clearly increasing and the delay in consulting patients is becoming shorter from year to year, as shown in the table and figure for the four years of the study.

Table LXXXI: Trend in time to thrombolysis over the four years of the study

Time to thrombolysis	Year 2019	Year 2020	Year 2021	Year 2022	TOTAL
< 2H	22	28	30	42	122
2 to 6H	37	34	32	34	137
> 6H	7	10	10	9	36
TOTAL	66	72	72	85	295
% Delay<2H	33%	39%	42%	49,4%	42%
% Delay 2-6H	56%	47%	44%	40%	45%
% Delay 6-12H	11%	14%	14%	10,6%	13%

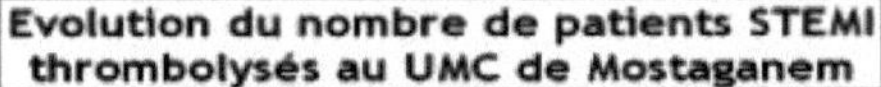

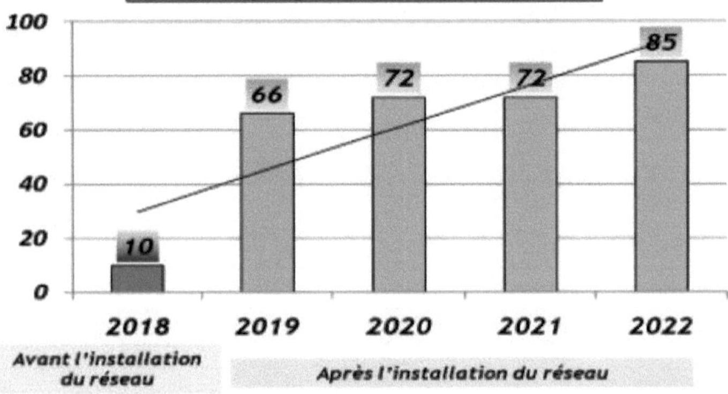

Figure 72: Trend in the number of patients treated with thrombolysis over the 4 years of the study

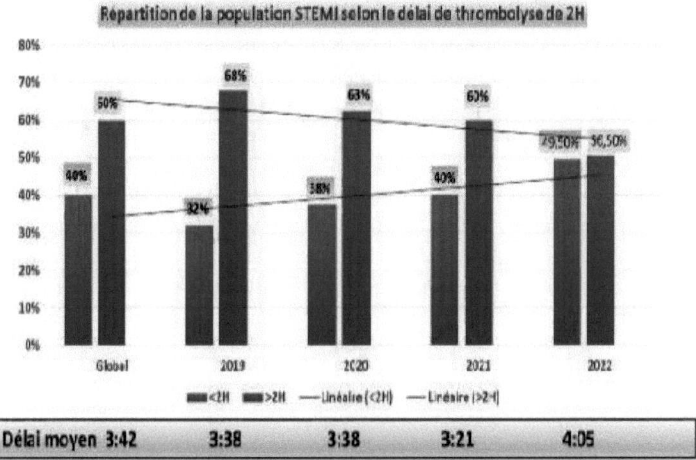

Figure 73: Distribution of STEMI patients according to 2-hour delay

16. Factors predictive of early STEMI complications on admission :

The predictive factors for complications of STEMI that were cross-referenced during our study, as shown in table 73, are mainly: age > 50 in men and > 60 in women, the existence of 2 or more CRDFs, admission of a patient in shock, a TIMI score > 6, a KILLIP grade > II and a HR > 100 beats per minute.

Table LXXXII: Factors predictive of early STEMI complications on admission

Death rates / Parameters	YES	NO	P
Class Men > 50	12	1	0,159
Classes Women > 60	6	1	0,479
Gender Male	12	8	0,003
BMI > 25	12	8	1
More than 2 FDRC	18	2	0,169
Existence of a Conduction Disorder	4	16	0,007
Existence of cardiogenic shock	14	6	0,0001
Delay > 2H	12	8	0,974
TIMI score > 6	17	3	0,000
KILLIP grade > 2	16	4	0,000
PAS100 < 100	11	9	0,000
HR >100	14	6	0,000
Anterior topography	9	11	0,034
Smoker	11	9	0,381
Diabetic	10	10	0 ,017
Hypertensive	12	8	0,027

17. Propose a thrombolysis strategy adapted to our region.

The primary aim of reperfusion therapy (whether using fibrinolytics or primary percutaneous coronary intervention (PCI)) is to restore blood flow to the previously viable ischaemic myocardium and reduce the size of the infarct.

Ongoing efforts are needed to improve pre-hospital care circuits (medical transport), to raise patients' awareness of the symptoms of STEMI and the need to seek medical advice as quickly as possible, and to equip Mostaganem with a catheterisation room in order to introduce PCI and improve the vital prognosis of STEMI patients in our region.
Thanks to the use of modern drug and interventional therapeutic strategies, it is possible to reduce mortality and morbidity in the early phase of myocardial infarction. To achieve this, we need to consider the need to :

- Helping GPs on duty on the outskirts of the CHU via teleconsultation (remote interpretation by a qualified doctor of the ECG taken in the field) to join our network and thrombolysing their patients on site, then sending them by medical ambulance to the CHU for ICU monitoring.
- Activate the role of the EMS in pre-hospital thrombolysis,
- Adopt an FDRC prevention strategy,
- Launch an information campaign aimed at the general public and the healthcare community about chest pain, in order to shorten the time it takes to decide to consult a doctor, and consequently the time it takes to get into hospital and receive treatment.

V.DISCUSSION :

Ischaemic heart disease is currently the leading cause of death worldwide.
Coronary pathology is currently the most common reason for emergency consultations.
In Algeria, mortality from myocardial infarction is not known with any precision. An increasing number of coronary syndromes are occurring in young people, with tragic consequences, with a national estimate of 24,000 ACS per year in Algeria [96].

A. Epidemiological characteristics :

1. Sex:

Our population of 295 coronary patients comprised 247 men **(83.7%)** and 48 women **(16.3%)**. There was a clear male predominance, with a sex ratio of **5.1**, i.e. five men for every woman.
This result is in line with the majority of studies which show that the frequency of coronary pathology in men is higher than in women, due to the fact that men consume more tobacco and other toxic products [294] and that women are protected until the menopause by restrogens.
A 2004 study at the Hussein-Dey University Hospital in Algiers found that 12% of 119 patients hospitalised for myocardial infarction were women [295].
In other studies in Algeria, a sex ratio of **3.87** was found in the multicentre STAMI study [296], **2.41** in the study by Boussouf et al [297] in Sétif and **2.4** in the study by Laraba [298] in Algiers. This male predominance has also been found in virtually all international studies. The sex ratio was **4.14** in Tunisia in the work by Milouchi et al [299], **3.87** in sub-Saharan Africa in the work by N'Guetta et al [300] and **3.22** in Egypt in the study by Gouda et al [301]. The rates of thrombolysis in women in the various international series fluctuate around 20%, as in the MONICA register [302] where the annual rates of myocardial infarction are 233/100,000 for men compared with 37/100,000 for women. In the GUSTO-1 study, out of a total of 41,021 patients included, only 25% were women.

Table LXXXIII: Comparison of the proportions of STEMI cases in the literature

Authors	McManus et al.	L. Belle et al.	N.Haouchine et al.	F. Addad et al.	H.Akoudad et al.	Our series

Study countries	United States	France	Algeria	Tunisia	Morocco	-
Reference Bibliograph.	[303]	[304]	[295]	[305]	[306]	-
Men (%)	60,6%	75%	88%	88,5%	75%	83,7 %
Women (%)	39,4%	25%	12%	11,5%	25%	16,3%

2. Age :

Age is a determining and aggravating factor in the accumulation of cardiovascular risk factors. The mean age of our sample of coronary patients was **57.85 ± 10.51 years**, with extremes ranging from 27 to 75 years.

The difference in mean age between the sexes is significant: men **(57.02 ± 10.52 years)** hospitalised in our study were younger than women **(62.13 ± 9.47 years)**.

The mean age of our results is in line with those reported by several international registries, such as the North African ACCESS registry, which is **59** years [307], the INTER HEART Africa study, which is **54.3** years [308], and the GULFRACE-2 registry, which is **56** years [309].

Table LXXXIV: Comparison of the mean age of onset of stemi in the literature

Authors	McManus et al.	L. Belle et al.	AM.Belgacem	F. Addad et al.	H.Akoudad et al.	Our series
Country	United States	France	Algeria	Tunisia	Morocco	-
Reference	[303]	[304]	[313]	[305]	[306]	-
Average age (years)	60 ± 13	63± 14	58.8 ± 10.3	60,8 ± 12,2	60	57,85 ± 10,51

In our study, up to the age of 55, men are the most frequently affected by STEMI; between the ages of 55 and 65, as many men as women are affected by STEMI, and over the age of 65, STEMI mainly affects women. This could be explained by the fact that women are protected until the menopause by restrogens.

The mean age of our patients (57.85 years) is younger than those included in the GRACE study [310] with a mean age of **64** years and in the CRUSADE study [311] with a mean age of **64.4** years.

In our study, the mean age of the women (62.13 years) was 5 years higher than that of the men (57.02 years), a finding found in several other series of studies such as GUSTO, where the women were on average 7 years older than the men **[266, 294, 312]**.

Age is a factor which limits the use of thrombolytics because of their haemorrhagic risk. This explains why most of the major thrombolysis trials prior to GUSTO imposed an upper age limit of 75 years.

Compared with several international series, our population is younger, with 86.4% (255 patients) aged less than 70 years on the day of the infarction, compared with only 13.6% (40 patients) aged between 70 and 75 years. In GISSI-2, 22.5% of the population studied were aged over 70 [264]; in ISIS-3, 26.1% of patients were aged 70 or over **[265]**.

In a study entitled "Management oriented towards reperfusion in the acute phase of myocardial infarction", carried out at the Hôpital Bichat [314] and involving 700 patients between 1988 and 1996, the mean age of the patients was 59 ± 13 years and 21% were

over 70 years of age.
Another Swiss study, also published in 1997 [315], collected 627 cases between 1986 and 1995 and found a mean patient age of 61 ±12 years (26 to 88 years), with 25% of patients aged over 70.
In Tunisia, in a prospective hospital study, without age limitation, carried out in 2005 in 5 cardiology departments [316], the mean age of men was 58.8 ± 12.2 years compared with 65.9 ± 11.5 years for women.

B. Coronary risk factors :

1. Diabetic population :

In our series, 27.1% of patients were known to be diabetic on the day of the infarction, 20.2% of them men and 62.5% women.

Table LXXXV: Percentages of diabetics+ stemi in the literature

STUDY	ISIS-3	GUSTO-1	GUSTO IIb	GUSTO III	Our series
Diabetics	**11.3**	**15**	**13.4**	**16**	**27,1**
Reference	[265]	[266]	[267]	[268]	-

82.5% of our diabetic patients are **aged over 55 and are** older than non-diabetics (mean age 62.26 ± **8.37 for diabetics compared with** 56.21 ± 10.77 for non-diabetics); this is probably due to the fact that almost half of them are women, who are older.
It is interesting to note that diabetes and overweight are strongly linked in our study, since the BMI (body mass index) of diabetic patients is higher than that of non-diabetics (26.83 ± 3.24 compared with 25.48 ± 3.37). This may be explained by the metabolic syndrome, in which diabetes and visceral obesity share the same pathogenesis, namely insulin resistance.

5. Hypertensive population :

The mean age of hypertensive patients was 62.44 ± 8.50 years, compared with 55.20 ± **10.66** years for non-hypertensive patients.
36.6% of our patients are confirmed hypertensives, and it is often diabetes that leads to the discovery of hypertension.
It should be noted that the combination of diabetes and hypertension affects 15.25% of our patients.

Table LXXXVI: Comparison of percentages of patients with hypertension + stemi in the literature

STUDIES	GABRIEL STEG	GUSTO-1	GUSTO IIb	GUSTO III	Our series
% HTA	**32,1**	**38**	**39**	**39,5**	**36.6**
Reference	[317]	[266]	[267]	[268]	-

Overall, our figures appear to be in line with those in the literature.
In GUSTO-I, nearly 38% of patients were hypertensive, compared with 39% and 39.5% respectively in GUSTO IIb and GUSTO III. In a study published in the JACC (Journal of the American college of cardiology), P.Gabriel Steg found that 32.1% of patients were hypertensive.

6. Smoking population :

The average age of smokers in our study was 55.46 ± **10.49** years**,** much lower than that of non-smokers, who were 62.12 ± **9.14** years.
64.1% of our patients were regular smokers who had not yet given up smoking on the day

of the infarction. This rate was 43% in GUSTO-I, 61% in GUSTO IIb and 41.4% in GUSTO III.

Table LXXXVII: Comparison of percentages of smokers + STEMI in the literature

STUDY	A.Bosset	GUSTO-1	GUSTO IIb	GUSTO III	Our series
of smokers	70	43	61	41,4	64,1
Reference	[318]	[266]	[267]	[268]	-

This figure of 64.1% is alarming, and calls for anti-smoking campaigns and the definitive elimination of smoking, as its cardiovascular risk persists for up to three years after stopping.

6. Dyslipidemic population :

The mean age of patients with dyslipidaemia was 57.89 ± **10.18** years, compared with 57.83 ± **10.73** years for those without dyslipidaemia.

37.3% of patients in our study had dyslipidaemia. The data in the literature on this subject remain fairly heterogeneous. In GUSTO III, between 34.4 and 34.8% of patients had hypercholesterolaemia; this rate was around 45% in STEG et al.

In an article by B. Thebault published in the Annales de Cardiologie et d'Angéiologie in 1997, 50% of patients aged under 75 had dyslipidaemia.

Patients' usual lipid profile is often incomplete, being limited to total cholesterol and triglyceride levels.

Table LXXXVIII: Comparison of the percentages of dyslipidemia + STEMI in the literature

STUDIES	G. STEG et al	B.Thebault	GUSTO III	Our series
Dyslipidemia	45%	50%	34.8%	37,3%
Reference	[317]	[319]	[268]	-

These figures bear witness to the impact of the new eating habits of our populations, which are undergoing a veritable epidemiological transition.

C.Infarct topography and haemodynamic status :

In our series, myocardial infarction was anterior in almost 2/3 of cases (58.1%).

In terms of haemodynamics, 2.3% of our patients were at Killip grade (III+IV) on admission. Our work shows that age greater than or equal to 65 years, diabetes and previous infarct topography are significantly associated with a worse Killip grade. These data were also found in the GUSTO-I study.

Table LXXXIX: Anterior location of infarction and killip grade

STUDIES	G. STEG et al	GISSI-2	GUSTO-1	GUSTO III	Our series
STEMI Anterior	49.9%	38%	37%	47.7%	58.1%
Killip III+IV		3.6%	13% (K>1)	2.1%	2.3%
Reference	[317]	[264]	[266]	[268]	-

The review of the literature shows quite different data; almost 38% of previous infarctions are recorded in GISSI-2 (3.6 to 4.1% of Killip III+IV patients), 49.9 to 51.2% for Steg et al, 47.7% in GUSTO -III (2.1% of Killip III+IV patients). It should be noted that in GUSTO-1 between 37% and 41% of patients had previous infarctions and between 13% and 17% had a Killip grade greater than I.

Haemodynamic status during myocardial infarction often depends on the extent and topography of necrosis, and it is now well established that the anterior location of the

infarct is a determinant of its morbidity and mortality [320].
There are currently mathematical models that can quantify the relative effects of different MI topographies, such as GUSTO, which shows that there are 7 extra years of age between the anterior and inferior locations.

D. Clinical characteristics

2. Consultation deadline :

Our study focused on treatment delays with the aim of identifying the parameters involved in the genesis of delays in consultation and management in order to try to correct them by planning educational and organisational interventions.
In our study, as in GUSTO, the times were divided into out-of-hospital and in-hospital times. The total time represents the time elapsed between the onset of pain and the start of thrombolysis. Out-of-hospital times include the time between the onset of symptoms and the patient's decision to consult a doctor, plus the time between the decision to consult a doctor and the first medical contact (FMC), then the time between the FMC and arrival at hospital. The intra-hospital delays include the diagnostic delays (validating ECG) and then the delays necessary for the thrombolysis decision from the moment the diagnosis is validated.
Of these phases, decision time is generally the main factor responsible for the overall delay.
A first observation in relation to the time of presentation of patients is that the distribution of our patients is not homogeneous between day and night, as we find that three quarters of our patients are hospitalised during the day (73.2%) compared with (26.8%) at night.
Analysis of factors influencing time to decision in STEMI patients was evaluated in Italy in the 1990s in a large multicentre case-control study of 5,301 patients in 118 coronary care units for the Survival in Infarction Study (GISSI) [321]. Among patient-related variables, older age, living alone, low initial symptom intensity, diabetes and nocturnal onset of symptoms appeared to significantly affect delay.
Balbaa et al [322] investigated the factors associated with a longer delay (>180 min) between symptom onset and primary PCI in STEMI patients admitted to two central cardiac centres in Egypt and Canada.
This observational study included all STEMI patients undergoing primary PCI over a one-year period at the tertiary heart centre in Aswan, Egypt (585 patients) and Hamilton General Hospital in Canada (715 patients). In Canada, there was no difference between men and women in the time of presentation to hospital, whereas in Egypt women had a longer delay between the onset of symptoms and the PCM. But the major difference between the two sites was in the mode of transport to hospital. In Canada, the majority of patients arrived at hospital in less than 20 minutes via the emergency medical system, whereas in Egypt the majority of patients took more than 40 minutes to reach hospital by taxi or public transport.
The average consultation time in our study was 3.85 hours from the onset of symptoms, reflecting the fact that patients are largely unaware of the importance of consulting a doctor early in the event of prolonged chest pain. This could lead to patients arriving at the emergency department in complicated conditions. These data should prompt greater efforts to raise public awareness of the possible symptoms of myocardial infarction.

3. Clinical signs on admission :

Chest pain has become one of the main complaints of patients visiting emergency departments [323]. It represents a major medical problem because it is a warning sign of cardiac ischaemia and infarction, if not pulmonary embolism, aortic dissection, compressive pneumothorax, pericardial tamponade or resophageal rupture. However, ACS accounts for only 20-25% of patients with chest pain presenting to emergency departments [324], and only 45% of patients admitted to a chest pain unit [325, 26]. Several other causes of chest pain have been recognised, including pneumonia, anxiety, migraine, musculoskeletal, skin and gastrointestinal disorders. In recent decades, several studies have shown a worryingly high rate (between 2% and 4%) of patients with undetected ACS being discharged from emergency departments [327, 328]. Unfortunately, wrongly discharged patients have a higher short-term mortality of between 10% and 25% [327, 328].

Today, practitioners still rely on Heberden's original description of angina pectoris, which he described as a sensation of strangulation radiating down the left arm [329].

Subsequently, the description of ischaemic chest pain has been progressively refined, as described in the recommendations of the National Heart Attack Alert Program Coordinating Committee [330], as pressure, tightness or heaviness. It may radiate to the neck, jaw, shoulders, back or one or both arms. The pain may also be described as indigestion or heartburn accompanied by nausea and/or vomiting.

Sweating associated with typical or atypical angina has also been shown, in a large cohort of over 12,000 ACS patients, to be a good predictor of ST-segment elevation myocardial infarction, rather than non-ST-segment elevation ACS, confirming its nature as a 'warning sign' for rapid assessment and treatment of patients with chest pain [331].

Another confounding factor is patient age. Although more than 80% of patients who die from coronary heart disease are over 65 years of age [332], the clinical presentation of ACS in the elderly is even more atypical (up to 50%), due to the number of associated comorbidities and complaints [333, 334].

Furthermore, it has been clearly demonstrated in Europe that elderly patients suffering from ACS have fewer ST-segment elevations and, despite their substantial in-hospital mortality, they receive less intensive treatment and investigation [335, 336].

In the Framingham study, which represents the largest and most studied group of patients in history, up to 25% of patients were found to have a Q-wave infarction on routine annual ECGs, in the absence of any previous clinical manifestation, which is a truly silent infarction [337].

A study of more than 434,800 patients included in the National Registry of Myocardial Infarction 2 (NRMI-2) database with a diagnosis of myocardial infarction found that 33% had no chest discomfort or arm, neck or jaw pain on initial presentation to hospital. MI patients without chest pain were, on average, seven years older and had a higher prevalence of women and diabetes than those with chest pain [338]. A Global Registry of Coronary Events (GRACE) study of over 20,000 ACS patients found that 8.4% did not have chest pain at presentation and that nearly a quarter were not diagnosed with ACS at initial assessment. The most common symptom in patients without chest pain was dyspnoea, followed by sweating, nausea and syncope. In this study, the absence of chest pain was also more common in the elderly, women, hypertensive patients, diabetics and patients with a history of congestive heart failure. Notably, mortality was much higher in all subgroups of patients without chest pain, being highest in patients with syncope [339]. Thus, the absence of chest pain appears to be one of the worst predictors of mortality.

In our sample, the inclusion criteria allowed us to select only patients presenting with a typical presentation, i.e. prolonged anginal pain evolving for less than 12 hours. Patients admitted to the emergency department for other reasons such as cardiogenic shock or syncope (as is the case in late-presenting cases) and then transferred to other departments where STEMI was diagnosed were not included in our study.

E. Support :

Primary percutaneous coronary intervention (or angioplasty) is currently the preferred reperfusion option for ST-segment elevation MI.

Balloon delay refers to the time interval between the arrival of a STEMI patient at the hospital and the time of balloon inflation (or stent deployment) that reopens the occluded artery or arteries. It forms part of the total time interval between the onset of occlusion and the reopening of the blocked artery or arteries (this time interval is commonly referred to as total ischaemic time (TIT)). Reducing this time shortens the ITT and is therefore considered an important strategy for achieving better patient outcomes. The idea of its importance was first suggested by the open artery theory [340], developed from animal experiments [341] and supported by angiographic studies in humans [342]. On the basis of this evidence, current practice guidelines [140] recommend a goal of 90 minutes for balloon-bearing delay as a strategy for achieving high-quality reperfusion. Following this recommendation, significant resources have been invested in reducing this time in healthcare systems around the world over the last few decades. The D2B Door-to-balloon Alliance, devised in 2006 by the American College of Cardiology, is perhaps the best known example.

When angioplasty cannot be performed within the recommended timeframe (either due to the absence of a catheterisation laboratory, unavailability of teams or transfer delays), the emergency physician should consider initiating immediate fibrinolytic therapy. It has been shown that administration of fibrinolytics within the first 2 hours after the onset of symptoms can reduce mortality by up to 48% [125].

Gersh et al [343] have suggested that the relationship between treatment delay, mortality and myocardial salvage can be represented by a curve, with the greatest benefit and sensitivity to delay residing during a critical early period of the first 2-3 hours after symptom onset.

Fibrinolytic therapy is more widely available than PCI and does not require an interventional cardiologist or catheterisation suite for administration. Rapid restoration of patency reduces infarct size, minimises the extent of myocardial damage, preserves left ventricular function and reduces morbidity [344].

As the time to reperfusion plays a key role in improving patient outcomes, a number of studies have evaluated the feasibility and efficacy of initiating fibrinolytic therapy before the patient arrives at hospital. The diagnosis of STEMI using 12-lead ECGs by emergency medical services has made it possible to initiate pharmacological reperfusion during transport to hospital. The clinical data collected to date are promising, but further studies are needed. A meta-analysis of data from studies of pre-hospital fibrinolysis demonstrated a 17% relative reduction in mortality rates with pre-hospital fibrinolysis compared with in-

hospital fibrinolysis [345].
The association between delay in ECP and mortality, as well as the development of heart failure, has been most clearly addressed by Therkelsen et al [346]. They showed that, for each hour of delay from the onset of symptoms to receiving treatment, the risk of death and the risk of heart failure increased by 10% over a three-year follow-up [347].

F. Next steps :

1. Complications :

The most common complication is **cardiogenic shock**, which affects around 10% of STEMI patients and is associated with a high risk of in-hospital mortality, generally greater than 40% [348].
The most common aetiology of cardiogenic shock is severe left ventricular (LV) dysfunction, but 10-20% of cases are secondary to mechanical complications or right ventricular infarction [349]. Patients with cardiogenic shock complicating STEMI are more likely to be elderly, to have had a previous stroke or transient ischaemic attack, to have a previous infarction or cardiac arrest on admission [350].
On the basis of the results of the physical examination, biomarkers and haemodynamics, **patients with STEMI and cardiogenic shock can be classified into five stages according to the "Society for Cardiovascular Angiography & Intervention"** (SCAI) **shock classification: Stage A**, at risk (includes all patients with STEMI); **Stage B**, early cardiogenic shock (haemodynamic instability or pre-shock); **Stage C**, classic cardiogenic shock; **Stage D**, deterioration; and **Stage E**, extreme [351]. The SCAI classification system has been validated in the cardiac intensive care unit at the Mayo Clinic, where a progressive increase in in-hospital mortality has been observed for each stage of SCAI shock [352].
Three therapeutic interventions should be carefully considered in patients with STEMI and cardiogenic shock, preferably as part of a multidisciplinary team: revascularisation (fibrinolysis/angioplasty/PAC), medical treatment (vasoactive/inotropic drugs) and/or mechanical circulatory support.
For patients with acute myocardial infarction and cardiogenic shock, urgent revascularisation is the standard of care [353].
Clinical practice guidelines recommend an early invasive strategy with appropriate revascularisation for all patients presenting with cardiogenic shock following MI [354], unless it is a case of refractory cardiogenic shock and clinical instability justifying immediate mechanical assistance.
The results we found in our study appear to be consistent with the data in the literature.
The patients who died in our sample were on average 6 years older (mean age 63.5 compared with 57.44) and included a higher proportion of women (16.66% compared with 4.85% of men).
Patients who died also had a greater history of diabetes, longer ischaemic times, unsuccessful thrombolysis and cardiogenic shock.
Our mortality rate was 6.78% in selected patients, which appears to be higher than the results of recent randomised trials. This may be explained by the non-selection of patients presenting to emergency with complicated conditions (cardiogenic shock, cardiac arrest) or those arriving late.
Cardiogenic shock was the leading cause of death in the acute phase after STEMI. It

seems sensible to pay more attention to strategies to prevent this complication, which may be best achieved by reducing total ischaemic time. Reducing the time from symptom onset to treatment has been shown to be crucial in reducing the incidence of death after STEMI [355].

In our study, the pain-thrombolysis time for STEMI patients improved after the installation of our network, but we believe that reducing pre-hospital time offers the greatest opportunity to improve prognosis after STEMI, and deserves greater attention and resource allocation.

Improving survival after STEMI can still benefit from several strategies which may include public education for early presentation, cardiac rehabilitation programmes and ongoing monitoring to encourage a healthy lifestyle and optimal medical management.

2. Mortality :

We can conclude that our patients who died were older, admitted late and in a more severe haemodynamic situation (thrombolysed in shock) when they were hospitalised.

It is also important to note that the mortality rate of STEMI patients is higher in women, as found by Otten et al [356] in a study of 6746 STEMI patients treated with primary angioplasty.

In order to explain this disparity, M. G. van der Meer et al [357] carried out a systematic review of gender differences in STEMI patients, including 21 studies that assessed the differences between women and men in baseline characteristics and short- and long-term outcomes, and it became clear that, overall, women were older and more often suffered from diabetes and hypertension. Remarkably, women had a longer delay between the onset of symptoms and receipt of care than men. Apparently, women tend to delay seeking medical care longer than men, confirming previous reports [358, 359].

Indeed, mortality is higher in women because of their unfavourable baseline risk profile; it is very important that clinicians recognise this difference and optimise their treatment and prevention strategies. This could lead to less disease and a better outcome, as cardiovascular disease remains the leading cause of death in women.

CHAPTER IV

CONCLUSION AND OUTLOOK

I. Conclusion :

Our work has clearly shown that setting up a thrombolysis network in the Mostaganem emergency department, by identifying and training the various players involved in this circuit, is a highly effective way of improving the management of patients suffering from ST-segment elevation acute coronary syndrome, both by reducing the various delays in diagnosis and by initiating treatment, thus providing an opportunity for rapid reperfusion with earlier medical surveillance and leading to a reduction in the morbidity and mortality of this serious public health condition.

The management of ST-segment elevation myocardial infarction (STEMI) is a race against time, since myocardial cell damage can occur after 20 to 30 minutes of ischaemia.

Reducing treatment time and maximising myocardial salvage in line with the mantra that "time is muscle" represents a logistical challenge.

Since the installation of our thrombolysis network in Mostaganem's emergency department, the time to thrombolysis has been an excellent measure of performance in terms of quality of care, aimed at speeding up the arrival of patients in the thrombolysis room.

However, there are currently a number of difficulties that can hamper the rapid management of STEMI patients and worsen their prognosis, in particular the inadequacy of medical transport, the lack of doctors and equipment in ambulances, the lack of an IT infrastructure enabling information to be transmitted and communication to take place between emergency doctors and paramedical teams in order to speed up the activation of catheterisation rooms, and the scarcity of centres capable of performing primary angioplasty. Most hospitals in Algeria capable of performing angioplasty do not have teams immediately available on site outside working hours, and the arrival of members of the interventional cardiology team is delayed in this context. As a result, STEMI patients who present out of hours are generally triaged to wait in A&E until the catheterisation team arrives and is ready to receive them.

Recently, more and more strategies have been implemented in the United States and Western Europe to minimise STEMI treatment times, including diagnosis by ECG interpretation in the pre-hospital setting, bypassing hospitals that do not have a cardiac catheterisation room, prior activation of catheterisation rooms by telephone call from the emergency doctor and even bypassing the emergency department.

In order to improve the quality of care and reduce delays in appropriate treatment, STEMI should be diagnosed early. With this in mind, the European recommendations updated in 2022 emphasise that emergency medical teams are responsible for the early diagnosis, triage, transport and treatment of patients with any ACS, including STEMI [360, 361].

Despite this ongoing quest to reduce treatment times, we must not overlook the importance of reducing the time between the onset of symptoms and the first medical contact, which is crucial to reducing total ischaemic time and improving therapeutic results. It is also time to look at developing systems that take care of patients from the onset of symptoms through to their return to normal life.

II. Future prospects :

Effective treatment of STEMI patients requires appropriate ambulance equipment and skills. All ambulances in the emergency medical system should be equipped with electrocardiogram (ECG) recording equipment, defibrillators and at least one person trained in resuscitation. All ambulance staff should be trained to recognise the clinical symptoms of myocardial infarction, to record and transmit the ECG, to administer oxygen if necessary, to relieve pain and to provide basic resuscitation care [362, 363] .

This strategy aims to reduce the time to treatment, leading to a reduction in mortality in STEMI patients. It allows immediate activation of the interventional team and direct transport of patients triaged for a primary PCI strategy to the catheterisation laboratory, bypassing the emergency department [362, 363].

Teleconsultation is a great help in this respect, allowing the transmission of the ECG performed in the field for possible remote interpretation by a qualified doctor, as well as the transmission of the patient's clinical data to the destination centre [362-364].

In addition to preliminary diagnosis and logistical aspects, teleconsultation should be used to coordinate pre-hospital treatment, particularly with regard to antiplatelet and anticoagulant therapy.

The European experience can serve as an example for improving these services, bypassing the emergency department in STEMI patients identified by pre-hospital ECG has been acquired as a strategy to optimise rapid reperfusion.

This approach has proved effective, with a 20-minute reduction in the median time between first medical contact and the balloon for patients bypassing A&E to go directly to the cardiac catheterisation room.

Despite the differences between our healthcare systems, this can be achieved in the coming years by implementing a number of measures, including integrating medical transport services into the national healthcare system, equipping ambulances with the equipment needed for the diagnosis and initial treatment of STEMI, training paramedical staff in the pre-hospital management of life-threatening situations, having a robust IT infrastructure enabling the digital transmission of pre-hospital ECGs, and setting up a STEMI network with 24-hour cardiac catheterisation centres.

It is predicted that over 70% of STEMI cases will occur in developing countries over the next ten years (428). This estimate highlights the urgent need to foster an effective system to promote and develop evidence-based revascularisation recommendations and educational initiatives for these countries.

The data from our study, which assesses the gap between daily practice and the recommendations, can be used as a basis for deploying more human and logistical resources to tackle this real public health problem and increase adherence to quality therapy.

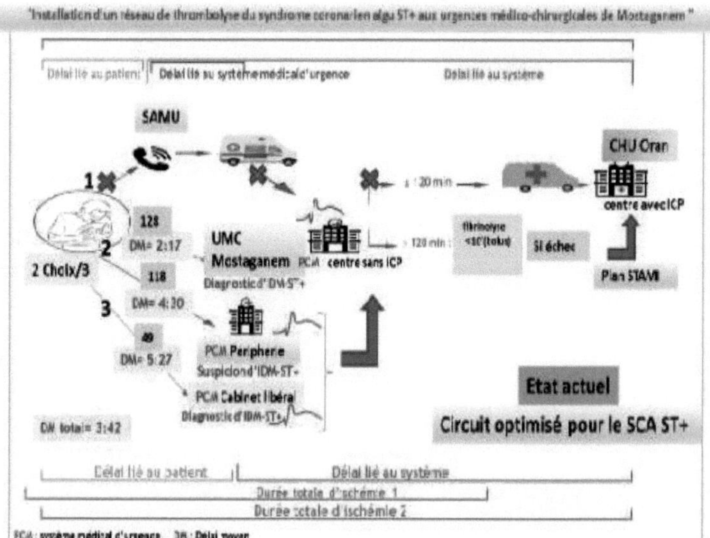

Figure 74: Current optimised STEMI patient circuit at Mostaganem University Hospital

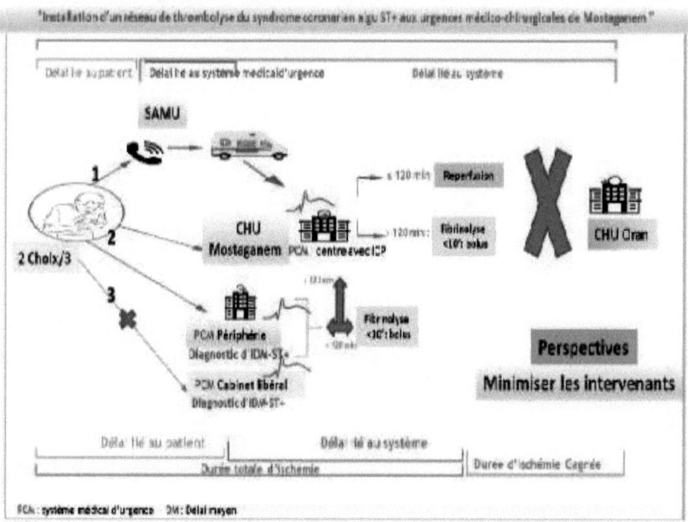

Figure 75: STEMI patient pathway at Mostaganem University Hospital

III. Declaration of interest :

No conflict of interest.

References

[1] . World Health Organization - the 10 leading causes of death in the world; fact sheets; January 2017.

[2] . Cardiovascular disease in europe 2014: epidemiological update m. nichols n. townsend p. scarborough m. rayner ***european heart journal***, volume 35, number 42, 7 november 2014, pages 2950-2959

[3] . Tahina project. analysis of causes of death in algeria 2002.

[4] . World health organization - no communicable diseases (ncd) country profiles, 2014.

[5] . World health organization - no communicable diseases (ncd) country profiles, 2018.

[6] . Management of acute coronary syndrome st + in algeria: dr benoui - pr daymellah univ. bejaia.

[7] . Haute autorité de santé - Management of acute myocardial infarction outside cardiology departments.

[8] . Libby p, pasterkamp g, crea f, jang ik. reassessing the mechanisms of acute coronary syndromes. circ res. 4 Jan 2019;124(1):150-60.

[9] . Thrombolysis past present and future d. gray postgrad med j 2006 82 372-375

[10] Giral P. Atheroma: pathological anatomy, physiopathology, epidemiology and risk factors, prevention. rev prat. 1998 ; 48 : 99-106.

[11] Cohen a. cardiologie et pathologie vasculaire. paris: estem, 1997.

[12] Akoudad h, benamer h. physiopathologie et étiopathogénie de l'infarctus du myocarde. emc-cardiologie angéiologie. 2004;1:49-67.

[13] (Santé publique france. Cardiovascular diseases. March 2016).

[14] (Joint international society and federation of cardiology; world health organization nomenclature and criteria for diagnosis of ischemic heart disease. report of the joint international society and federation of cardiology/world health organization task force on standardization of clinical nomenclature. circulation. 1979 mar;59(3):607-9.).

[15] (European society of cardiology; american college of cardiology committee. myocardial infarction redefined a consensus document of the joint european society of cardiology/american college of cardiology committee for the redefinition of myocardial infarction. eur heart j. 2000 sep; 21(18):1502-13.)

[16] (Drake rl, vogl aw, mitchell awm. gray's anatomy for students. 2nd edition. paris: elsevier; 2010).

[17] The fundamentals of cardiovascular pathology integrated teaching - cardiovascular system. elsevier masson. 2014. p.24-25. isbn: 9782294721137).

[18] (Collège national des enseignants de cardiologie; société française de cardiologie. les fondamentaux de la pathologie cardiovasculaire enseignement intégré - système cardiovasculaire. elsevier masson. 2014. p.24-25. isbn: 9782294721137).

[19] (Drake rl, vogl aw, mitchell awm. gray's anatomy for students. 2nd edition. paris: elsevier; 2010).

[20] (Collège national des enseignants de cardiologie; société française de cardiologie. atheroma: epidemiology and pathophysiology. the polyatheromatous patient. cardiology. second edition. elsevier masson. 2015.

[20]Bauters c. Physiopathology of atherosclerosis. available at http://www.pifo.uvsq.fr/hebergement/cec_mv/128b.pdf):

[21] (Duriez p. mécanismes de formation de la plaque d'athérome. rev med inter

2004;25suppl1:s3-6)
[22] (Atherosclerosis: prevention of vascular risks . available at https://cprv.pagesperso-orange.fr/atherome)
[23] (Santré c. acute coronary syndromes. February 2002. available at http://reaannecy.free.fr/documents/cardiologie/syndromes_coronariens_aigus.htm).
[24] -Stanley wc et al.regulation of myocardial carbohydrate metabolism under normal and ischaemic conditions . cardiovasc res.1997; 33:243-57.
[25] -Liu et al. circ res. 1996 ; 79 :940-8.
[26]-Deutsch e; berger m; kusmaull wg et al.adapttion to ischemia during percutaneous transluminal coronary angioplasty. clinical hemodynamic and metabolic features. circulation 1990; 82:2044-51.
[27]-Ovize m; henry p; rioufol g; minaire y. le préconditionnement ischémique: concept of endogenous myocardial protection. arch mal creur 1995 ;88 : 869-77.
[28]-St John sutton mg; sharp n. left ventricular remodeling after myocardial infarction.pathophysiology and therapy. circulation2000;101:2981-2988.
[29]- Anversa p; olivetti g; capasso j.cellular basis of ventricular remodeling after myocardial infarction. am j cardiol 1991;68:7-16.
[30]-Bonnefoy e, grollier g, fradin s, scanu p, tessier p, valette b, foucault jp, potier jc. critères précoces, cliniques, électrocardiographiques, et biologiques de reperfusion après fibrinolyse intraveineuse à la phase aiguê de l'infractus du myocarde. arch mal creur 1993; 86: 857- 63.
[31]-Boli r, marban e. molecular and cellular mechanisms of myocardial stunning. physiol rev 1999; 79:609-634.
[32]-Braunwald e; kloner ra. the stunned myocardium: prolonged post ischemic ventricular dysfunction. circulation 1982; 66:1146-9.
[33]-Rahimtoola sh.from coronary artery disease to heart failure: role of the hibernating myocardium. am j cardiol 1995;75:16e-22e.
[34]. Hammer a. ein fall von thrombotischem verschlusse einer der kranzarterien des herzens. wien med wschr. 1878;28:97-102.
[35]. Obraztzow vp, straschesko nd. zur kenntnis der thrombose der koronararterien des herzens. z klin med. 1910;71:116-132.
[36]. Herrick jb. clinical features of sudden obstruction of the coronary arteries. jama.1912;59:2015-2022.
[37] Herrick jb. clinical features of sudden obstruction of the coronary arteries. *jama*, 1912; 59: 2 015-2 020.]
[38] - World health organisation: nomenclature and criteria for diagnosis of ischaemic heart disease. report of the joint international society and federation of cardiology/ world heart organisation task force on standardisation of clinical nomenclature. circulation 1979; 59: 607-609.
[39]. Nomenclature and criteria for diagnosis of ischemic heart disease. report of the joint international society and federation of cardiology/world health organization task force on standardization of clinical nomenclature. circulation 1979 ; 59 : 607-9.
[40] (The joint european society of cardiology/ american college of cardiology committee . myocardial infarction redefined- a consensus document of the joint european society of cardiology/ american college of cardiology committee for the redefinition of myocardial infarction . eur heart j 2000;21(18):1502-1513.)

[41] Myocardial infarction redefined--a consensus document of the joint european society of cardiology/american college of cardiology committee for the redefinition of myocardial infarction. eur heart j 2000; 21: 1502-13)
[42] (Thygesen k, alpert js, white hd; joint esc/acc/aha/whf task force for the redefinition of myocardial infarction. universal definition of myocardial infarction. eur heart j. 2007;28:2525-2538; circulation. 2007;116:2634-2653; j am coll cardiol. 2007;50:2173-2195.).
[43]. (The joint european society of cardiology/american college of cardiology committee. myocardial infarction redefined-a consensus document of the joint european society of cardiology/american college of cardiology committee for the redefinition of myocardial infarction. eur heart j. 2000;21:1502-1513; j am coll cardiol. 2000;36:959- 969).
[44]. (Thygesen, k. et al. fourth universal definition of myocardial infarction.)
[45]. Sandoval y, thygesen k. myocardial infarction type 2 and myocardial injury. clin chem. 2017;63:101-107.
[46]. Williams rr, hunt sc, heiss g, province ma, bensen jt, higgins m, et al. usefulness of cardiovascular family history data for population-based preventive medicine and medical research (the health family tree study and the nhlbi family heart study). am j cardiol.2001;87(2):129-35.
[47]. (World Health Organization. tobacco. key facts. available at: https://www.who.int/fr/news-room/fact-sheets/detail/tobacco)
[48]. (Ambrose ja, barua rs. the pathophysiology of cigarette smoking and cardiovascular disease. j am
coll cardiol. 2004;43(10):1731-1737).
[49]. (Collège national des enseignants de cardiologie et de la société française de cardiologie. cardiologie. elsevier masson s.a.s. 470694 - (i) - (3) - csb80 - noc. 2010).
[50]. World Health Organization. The WHO is calling on countries to reduce the sugar intake of adults and children. available on:https://www.who.int/mediacentre/news/releases/2015/sugar-guideline/fr/
[51].World Health Organization. reduce salt intake. key facts. june 2016 available at : http://www.who.int/fr/news- room/fact-sheets/detail/salt- reduction
[52]. Ministère de la santé, de la population et de la réforme hospitalière ; organisation mondiale de la santé. mesure des facteurs de risque des maladies non transmissibles dans deux wilayas pilotes en algérie (approche step "wise" de l'oms). rapport final 2005.
[53]. World Health Organization. Global Strategy on Diet, Physical Activity and Health. available at http://www.who.int/dietphysicalactivity/pa/fr/
[54].World Health Organization. health topics, obesity. available at https://www.who.int/topics/obesity/fr/
[55].Bodenant m, kuulasmaa k, wagner a, kee f, palmieri l, ferrario mm, et al; morgam project. measures of abdominal adiposity and the risk of stroke: the monica risk, genetics, archiving and monograph (morgam) study. stroke. 2011 oct;42(10):2872-7.
[56] Von känel r, mills pj, fainman c, dimsdale je. effects of psychological stress and psychiatric disorders on blood coagulation and fibrinolysis: a biobehavioral pathway to coronary artery disease? psychosom med. 2001 jul- aug;63(4):531-44.
[57].Fédération française de cardiologie. creur et stress. combattre le stresspour réduire le risque cardiovasculaire. paris. available on : www.fedecardio.com
[58].Domínguez f, fuster v, fernández-alvira jm, fernández-friera l, lópez-melgar b,

blanco-rojo r, et al. association of sleep duration and quality with subclinical atherosclerosis. j am coll cardiol. 2019 jan 22;73(2):134-144.
[59].Jones rt. cardiovascular system effects of marijuana. j clin pharmacol 2002;42(suppl 11):58-63.
[60] Mittleman ma, lewis ra, maclure m, sherwood jb, muller je. triggering myocardial infarction by marijuana. circulation 2001; 103:2805-9.
[61] Agence française de sécurité sanitaire des produits de santé. prise en charge thérapeutique du patient dyslipidémique. recommandations. saint- denis: afssaps; 2005.
[62].Yadav as, bhagwat vr. lipid profile pattern in anginal syndrome patients from marathwada region of maharashtra state. journal of medical education & research. 2012;2(2):12-15.
[63] Lecoffre c, perrine al, blacher j, olié v. cholestérol ldl chez les adultes en france métropolitaine : concentration moyenne, connaissance et traitement en 2015, évolutions depuis 2006. bull epidémiol hebd. 2018; (37):710-8.
[64] Agmon y, khandheria bk, meissner i, schwartz gl, petterson tm, o'fallon wm, gentile f, whisnant jp, wiebers do, seward jb. independent association of high blood pressure and aortic atherosclerosis: a population- based study. circulation. 2000 oct 24; 102(17):2087-93.
[65].Sen S, oppenheimer sm, lima j, cohen b. risk factors for progression of aortic atheroma in stroke and transient ischemic attack patients. stroke. 2002 apr;33(4):930-5.
[66].Milouchi s, ajmi h, sghaier a, khorchani a, ferjani s. management of acute coronary syndrome with st segment elevation in medenine regional hospital. a propos de 150 cas. cardiologie tunisienne.2eme trimestre 2017; 13(02) :93-100.
[67].Zouzou h. determination of the frequency and study of the predictive factors of rhythm and conduction disorders during acute coronary syndrome with elevation of the st segment during the hospital phase [thesis]. alger: université d'alger 1; 2016.
[68] Charfeddine s, ellouze t, abid l, hammami r, hamza c, maalej a, et al. management of myocardial infarction with persistent st segment elevation: prospective study about 215 patients. cardiologie tunisienne. 4th quarter 2017; 13(04):11-18.
[69] Passeron j. guide pratique des facteurs de risque cardio-vasculaire. paris: mmi éditions, 2000, 248 p.
[70].World health organization. use of glycated haemoglobin (hba1c) in the diagnosis of diabetesmellitus , 2011. disponiblesur: http://www.who.int/diabetes/publications/diagnosis_diabetes2011/en/
[71].American diabetes association. classification and diagnosis of diabetes: standards of medical care in diabetes. 2018 diabetes care. 2018;41(suppl. 1):s13-s27
[72] Norhammar a, tenerz a, nilsson g, hamsten a, efendíc s, rydén l, et al. glucose metabolism in patients with acute myocardial infarction and no previous diagnosis of diabetes mellitus: a prospective study.lancet. 2002 jun22; 359(9324):2140-4.
[73].Xavier g. la prévention des maladies cardiovasculaires. adsp. juin 2004; 47: 47.
[74] Vergès b, zeller m, dentan g, beer jc, laurent y, janin-manificat l, et al. impact of fasting glycemia on short-term prognosis after acute myocardial infarction. j clin endocrinol metab. 2007 jun;92(6):2136-40.
[75] Lee wl, cheung am, cape d, zinman b. impact of diabetes on coronary artery disease in women and men: a meta-analysis of prospective studies. diabetes care. 2000 jul;23(7):962-8.

[76] Eckel rh, grundy sm, zimmet pz. the metabolic syndrome. lancet.2005; 365: 141528.
[77] Shin ja, lee jh, lim sy, ha hs, kwon hs, park ym, et al. metabolic syndrome as a predictor of type 2 diabetes, and its clinical interpretations and usefulness. j diabetes investig. 2013 jul 8;4(4):334-43.
[78] .International diabetes federation,. professors sir george alberti and paul zimmet. the idf consensus worldwide definition of the metabolic syndrome. 2006.
[79] Isomaa b, almgren p, tuomi t, forsén b, lahti k, nissén m, et al. cardiovascular morbidity and mortality associated with the metabolic syndrome. diabetes care. 2001 apr; 24(4):683-9.
[80] Stengel b, billon s, van dijk pcw, jager kj, dekker fw, simpson k, et al. trends in the incidence of renal replacement therapy for end-stage renal disease in europe, 1990-1999. nephrol dial transplant. 2003;18:1824-33.
[81] .Us renal data system: usrds. annual data report. bethesda, mdm national institutes of health, national institute of diabetes and digestive and kidney diseases. 2003.
[82] .J, selvin e, stevens la, manzi j, kusek jw, eggers p, et al. prevalence of chronic kidney disease in the united states. jama 2007;298:2038-47.
[83] Fesler p, ribstein j. moderate impairment of renal function and cardiovascular risk. rev med interne. 2009;30:585-91.
[84] .Go a, chertow g, fan d, mcculloch ce, hsu cy. chronic kidney disease and the risks of death, cardiovascular events, and hospitalization. n englj med.2004;351:1296-305.
[85] .World health organisation. the world health report 2014.geneva: who. 2014
[86] .World Health Organization. cardiovascular disease. available at : http://www.who.int/mediacentre/factsheets/fs317/fr.
[87] .World Health Organization. 10 leading causes of death. available at : http://www.who.int/mediacentre/factsheets/fs310/fr/ updated24 May 2018.
[88] Nichols m, townsend n, scarborough p, et al. cardiovascular disease in europe 2014: epidemiological update. eur heart j 2014;35:2950-9.
[89] .Tunstall-pedoe h, kuulasmaa k, mähönen m, tolonen h, ruokokoski e, amouyel p. contribution of trends in survival and coronary-event rates to changes in coronary heart disease mortality: 10-year results from 37 who monica project populations. monitoring trends and determinants in cardiovascular disease. lancet.1999; 353(9164): 1547-1557.
[90] Gabet a, lamarche vadel a, chin f, olie v. disparités régionales de la mortalité prématurée par maladie cardiovasculaire en france (2008-2010) et évolutions depuis 2000-2002. bulletin épidémiologique hebdomadaire, 2014; 26:430-438
[91] .Ministère de la santé, de la population et de la réforme hospitalière, institut national de santé publique. analyse des causes de décès, année 2002. projet tahina. novembre 2008.
[92] .Institut national de santé publique, service des causes médicales des décès. problématique du certificat médical de décès en algérie. information day on the medical death certificate 2017.
[93] .Bruno p, régis b, francois c et al .therapeutic education in coronary heart disease: position paper from the working group of exercise rehabilitation and sport (gers) and the therapeutic education commission of the french society of cardiology. archives of cardiovascular disease.2013; 106 :680- 89.
[94] .Zouzou h. determination of the frequency and study of the predictive factors of rhythm and conduction disorders during acute coronary syndrome with elevation of the st segment during the hospital phase [thesis]. alger: université d'alger 1; 2016.

[95] Baudin b, cohen a, berthelot-garcias e, meuleman c, dufaitre g, ederhy s, et al. epidemiological data on cardiovascular disease and management of cardiovascular events. rev fr labo. 2009;39(409):27-39.

[96] Steg pg, goldberg rj, gore jm. baseline characteristics, management practices, and inhospital outcomes of patients hospitalized with acute coronary syndromes in the global registry of acute coronary events (grace). acc current j review. 2002;11(6):16-7.

[97].Gach o, el husseini z, lancellotti p. acute coronary syndrome. rev med liege. 2018;73(5-6):243-50.

[98] N'guetta r, yao h, ekou a, n'cho-mottoh mp, angoran i, tano m, et al. prevalence and characteristics of acute coronary syndromes in a sub-Saharan african population. ann cardiol angéiol. 2016;65(2):59-63.

[99] Hunziker p, pfisterer m, marsch s. acute coronary syndrome: diagnosis and risk stratification. forum med suisse. 2003;(25):580-4.

[100] .Grech ed. pathophysiology and investigation of coronary artery disease. bmj 2003;326:1027-30.

[101] .wagner gs, macfarlane p, wellens h, et al. aha/accf/hrs recommendations for the standardization and interpretation of the electrocardiogram: part vi: acute ischemia/infarction: a scientific statement from the american heart association electrocardiography and arrhythmias committee. council on clinical cardiology; the american college of cardiology foun- dation; and the heart rhythm society: endorsed by the international society for computerized electrocardiology. cir- culation 2009;119:e262-70.

[102] Sgarbossa eb, pinski sl, barbagelata a, et al. electrocardio- graphic diagnosis of evolving acute myocardial infarction in the presence of left bundle-branch block. global utilization of streptokinase and tissue plasminogen activator for occlu- ded coronary arteries investigators (gusto-1). n engl j med 1996;334:481-7.

[103] Sgarbossa eb, pinski sl, gates kb, wagner gs. early electro- cardiographic diagnosis of acute myocardial infarction in the presence of ventricular paced rhythm. gusto-i investigators. am j cardiol 1996;77:423-4.

[104] Das mk, khan b, jacob s, kumar a, mahenthiran j. signi- ficance of a fragmented qrs complex versus a q wave in patients with coronary artery disease. circulation 2006;113: 2495-501.

[105] .de winter rj, verouden nj, wellens hj, wilde aa. a new ecg sign of proximal lad occlusion. n engl j med 2008;359: 2071-3.

[106] .antman em, cohen m, bernink pj, et al. the timi risk score for unstable angina/non-st elevation mi: a method for prognostication and therapeutic decisionmaking. jama 2000;284:835-42.

[107] .Bassand jp, hamm cw, ardissino d, et al. guidelines for the diagnosis and treatment of non-st-segment elevation acute coronary syndromes. eur heart j 2007;28:1598-660.

[108] Eagle ka, lim mj, dabbous oh, et al. a validated prediction model for all forms of acute coronary syndrome: estimating the risk of 6-month postdischarge death in an international registry. jama 2004;291:2727-33.

[109] Cannon cp, weintraub ws, demopoulos la, et al. comparison of early invasive and conservative strategies in patients with uns- table coronary syndromes treated with the glycoprotein iib/iiia inhibitor tirofiban. n engl j med 2001;344:1879-87.

[110] Mehta sr, granger cb, boden we, et al. early versus delayed invasive intervention in acute coronary syndromes. n engl j med 2009;360:2165-75.

[111] Abou tam j, buffet p, lorgis l, zeller m, gonzalez s, F'huillier i, et al. risk stratification scores and acute coronary syndromes. annales de cardiologie et d "angéiologie. august 2005;54(4):157- 60.

[112] .authors/task force members, hamm cw, bassand j-p, agewall s, bax j, boersma e, et al. esc guidelines for the management of acute coronary syndromes in patients presenting without persistent st-segment elevation: the task force for the management of acute coronary syndromes (acs) in patients presenting without persistent st-segment elevation of the european society of cardiology (esc). european heart journal. 26 august 2011;32(23):2999- 3054.

[113] .Morrow da, antman em, charlesworth a, cairns r, murphy sa, lemos ja de, et al. timi risk score for st-elevation myocardial infarction: a convenient, bedside, clinical score for risk assessment at presentation : an intravenous npa for treatment of infarcting myocardium early ii trial substudy. circulation. 24 oct 2000;102(17):2031- 7.

[114] .Ibanez b, james s, agewall s, antunes mj, bucciarelli-ducci c, bueno h, et al. 2017 esc guidelines for the management of acute myocardial infarctionin patients presenting with st-segment elevation: the task force for the management of acute myocardial infarction in patients presenting with st- segment elevation of the european society of cardiology (esc). eur heart j. 2018 jan 7;39(2):119-177.

[115] Percutaneous coronary intervention . available on: http://www.nhlbi.nih.gov/health/dci/diseases/angioplasty/angioplasty_whatis. html

[116] Review of biomechanical studies of arteries and their effect on stent performance. available on: http://www.medivisuals.com/an-arterial-wall- 604089r-02x.aspx.

[117] Zijlstra f, hoorntje jc, boer mj, reiffers s, miedema k, ottervanger jp, et al. longterm benefit of primary angioplasty as compared with thrombolytic therapy for acute myocardial infarction. n engl j med.1999; 341:1413-1419.

[118] .Keeley ec, boura ja, grines cl. primary angioplasty vs. intravenous thrombolytic therapy for acute myocardial infarction: a quantitative review of 23 randomised trials. lancet 2003;361:13-20.

[119] Widimsky p, budesinsky t, vorac d, groch l, zelizko m, aschermann m, et al. long distance transport for primary angioplasty vs. immediate thrombolysis in acute myocardial infarction. final results of the randomized national multicentre triale prague- 2. eur heart j. 2003;24:94-104.

[120] Andersen hr, nielsen tt, rasmussen k, thuesen l, kelbaek h, thayssen p, et al. a comparison of coronary angioplasty with fibrinolytic therapy in acute myocardial infarction. n engl j med. 2003;349:733-742.

[121] Terkelsen cj, lassen jf, norgaard bl, gerdes jc, jensen t, gotzsche lb, et al. mortality rates in patients with st-elevation vs. non-st-elevation acute myocardial infarction: observations from an unselected cohort. eur heart j 2005; 26:18-26.

[122] .E.does time matter? a pooled analysis of randomized clinical trials comparing primary percutaneous coronary intervention and in-hospital fibrinolysis in acute myocardial infarction patients. eur heart j 2006; 27:779- 788.

[123] Pinto ds, frederick pd, chakrabarti ak, kirtane aj, ullman e, dejam a, miller dp, et al. benefit of transferring st-segment-elevation myocardial infarction patients for percutaneous coronary intervention compared with administration of onsite fibrinolytic declines as delays increase. circulation. 2011;124:2512-2521.

[124] .Jp t, ap t. a review of available fibrin-specific thrombolytic agents used in acute myocardial infarction. pharmacotherapy [internet]. feb 2001 [cited 10 Sep 2022];21(2). available from: https://pubmed.ncbi.nlm.nih.gov/11213858/

[125] Morrison lj, verbeek pr, mcdonald ac, sawadsky bv, cook dj. mortality and prehospital thrombolysis for acute myocardial infarction. jama 2000;283:2686-92.

[126] .Boersma e, maas ac, deckers jw, simoons ml. early thrombolytic treatment in acute myocardial infarction: reappraisal of the golden hour. lancet 1996;348:771-5.

[127] Rawles jm, kenmure ac. controlled trial of oxygen in uncomplicated myocardial infarction. br med j. 8 may 1976;1(6018):1121-3.

[128] . Steg pg, bonnefoy e, chabaud s, lapostolle f, dubien py, cristofini p, et al. impact of time to treatment on mortality after prehospital fibrinolysis or primary angioplasty: data from the captim randomized clinical trial. circulation. 9 Dec 2003;108(23):2851-6.

[129] Ibanez, b. et al. 2017 esc guidelines for the management of acute myocardial infarction in patients presenting with st-segment elevation: the task force for the management of acute myocardial infarction in patients presenting with st-segment elevation of the european society of cardiology (esc).

[130] Danchin n, coste p, ferrières j, steg pg, cottin y, blanchard d, et al. comparison of thrombolysis followed by broad use of percutaneous coronary intervention with primary percutaneous coronary intervention for st-segment-elevation acute myocardial infarction: data from the french registry on acute st-elevation myocardial infarction (fast-mi). circulation. 15 Jul 2008;118(3):268-76.

[131] Gershlick ah, stephens-lloyd a, hughes s, abrams kr, stevens se, uren ng, de belder a, davis j, pitt m, banning a, baumbach a, shiu mf, schofield p, dawkins kd, henderson ra, oldroyd kg, wilcox r, react trial investigators. rescue angioplasty after failed thrombolytic therapy for acute myocardial infarction.

[132] Cantor wj, fitchett d, borgundvaag b, ducas j, heffernan m, cohen ea, et al. routine early angioplasty after fibrinolysis for acute myocardial infarction. n engl j med. 25 june 2009;360(26):2705-18.

[133] . Di mario c, dudek d, piscione f, mielecki w, savonitto s, murena e, et al. immediate angioplasty versus standard therapy with rescue angioplasty afterthrombolysis in the combined abciximab reteplase stent study in acute myocardial infarction (caressin-ami): an open, prospective, randomised, multicentre trial. lancet. 16 Feb 2008;371(9612):559-68.

[134] . B0hmer e, hoffmann p, abdelnoor m, arnesen h, halvorsen s. efficacy and safety of immediate angioplasty versus ischemia-guided management after thrombolysis in acute myocardial infarction in areas with very long transfer distances results of the nordistemi (norwegian study on district treatment of st-elevation myocardial infarction). j am coll cardiol. 12 jan 2010;55(2):102-10.

[135] . Le May mr, wells ga, labinaz m, davies rf, turek m, leddy d, et al. combined angioplasty and pharmacological intervention versus thrombolysis alone in acute myocardial infarction (capital ami study). j am coll cardiol. 2 august 2005;46(3):417-24.

[136] . Fernandez-avilés f, alonso jj, castro-beiras a, vázquez n, blanco j, alonso-brialesj, et al. routine invasive strategy within 24 hours of thrombolysis versus ischaemia- guided conservative approach for acute myocardial infarction with st-segment elevation (gracia-1): a randomised controlled trial. lancet. 18 sept 2004;364(9439):1045-53.

[137] . Borgia f, goodman sg, halvorsen s, cantor wj, piscione f, le may mr, et al. early routine percutaneous coronary intervention after fibrinolysis vs. standard therapy in st-segment elevation myocardial infarction: a meta-analysis. eur heart j. sept 2010;31(17):2156-69.

[138] . Bonnefoy e, steg pg, boutitie f, dubien py, lapostolle f, roncalli j, et al. comparison of primary angioplasty and pre-hospital fibrinolysis in acute myocardial infarction (captim) trial: a 5-year follow-up. eur heart j. juill 2009;30(13):1598-606.

[139] . D'souza sp, mamas ma, fraser dg, fath-ordoubadi f. routine early coronary angioplasty versus ischaemia-guided angioplasty after thrombolysis in acute st- elevation myocardial infarction: a meta-analysis. eur heart j. apr 2011;32(8):972-82.

[140] . O'gara pt, kushner fg, ascheim dd, casey de, chung mk, de lemos ja, et al. 2013 accf/aha guideline for the management of st-elevation myocardial infarction: a report of the american college of cardiology foundation/american heart association task force on practice guidelines. j am coll cardiol. 29 jan 2013;61(4):e78-140.

[141] . Yusuf s, mehta sr, chrolavicius s, afzal r, pogue j, granger cb, budaj a, peters rjg, bassand jp, wallentin l, joyner c, fox kaa, oasis-6 trial group. effects of fondaparinux on mortality and reinfarction in patients with acute st-segment elevation myocardial infarction: the oasis-6 randomized trial - pubmed. available on: https://pubmed.ncbi.nlm.nih.gov/16537725/

[142] . Montalescot g, zeymer u, silvain j, boulanger b, cohen m, goldstein p, et al. intravenous enoxaparin or unfractionated heparin in primary percutaneous coronary intervention for st-elevation myocardial infarction: the international randomised openlabel atoll trial. lancet. 20 august 2011;378(9792):693-703.

[143] . Jp c, k h, m c, u z, p g, jr pc, et al. a direct comparison of intravenous enoxaparin with unfractionated heparin in primary percutaneous coronary intervention (from the atoll trial). am j cardiol. 5 Sep 2013;112(9):1367-72.

[144] . J s, f b, o b, c p, m c, u z, et al. efficacy and safety of enoxaparin versus unfractionated heparin during percutaneous coronary intervention: systematic review and meta-analysis. bmj (clinical research ed) [internet]. 2 March 2012;344.
available at: https://pubmed.ncbi.nlm.nih.gov/22306479/

[145] . Mega jl, braunwald e, wiviott sd, bassand jp, bhatt dl, bode c, et al. rivaroxaban in patients with a recent acute coronary syndrome. n engl j med. 5 jan 2012;366(1):9-19.

[146] . Damman p, woudstra p, kuijt wj, de winter rj, james sk. p2y12 platelet inhibition in clinical practice. j thromb thrombolysis. feb 2012;33(2):143-53.

[147] . Wiviott sd, steg pg. clinical evidence for oral antiplatelet therapy in acute coronary syndromes. lancet. 18 Jul 2015;386(9990):292-302.

[148] . Levine gn, bates er, bittl ja, brindis rg, fihn sd, fleisher la, et al. 2016 acc/aha guideline focused update on duration of dual antiplatelet therapy in patients with coronary artery disease: a report of the american college of cardiology/american heart association task force on clinical practice guidelines: an update of the 2011 accf/aha/scai guideline for percutaneous coronary intervention, 2011 accf/aha guideline for coronary artery bypass graft surgery, 2012 acc/aha/acp/aats/pcna/scai/sts guideline for the diagnosis and management of patients with stable ischemic heart disease, 2013 accf/aha guideline for the management of st-elevation myocardial infarction, 2014 aha/acc guideline for the management of patients with non-st-elevation acute coronary syndromes, and 2014 acc/aha guideline on perioperative cardiovascular evaluation and management of patients undergoing noncardiac surgery. circulation. 6 Sep
2016;134(10):e123-155.

[149] . Valgimigli m, bueno h, byrne ra, collet jp, costa f, jeppsson a, et al. 2017 esc focused update on dual antiplatelet therapy in coronary artery disease developed in collaboration with eacts: the task force for dual antiplatelet therapy in coronary artery disease of the european society of cardiology (esc) and of the european association for cardio-thoracic surgery (eacts). eur heart j. 14 jan 2018;39(3):213-60.

[150] . Zeymer u, hohlfeld t, vom dahl j, erbel r, münzel t, zahn r, et al. prospective, randomised trial of the time dependent antiplatelet effects of 500 mg and 250 mg acetylsalicylic acid i. v. and 300 mg p. o. in acs (acute). thromb haemost. 28 feb 2017;117(3):625-35.

[151] . G m, aw van 't h, f l, j s, jf l, l b, et al. prehospital ticagrelor in st-segment elevation myocardial infarction. the new england journal of medicine [internet]. 9 Nov 2014 ;371(11). available at: https://pubmed.ncbi.nlm.nih.gov/25175921/

[152] . Effect of upstream clopidogrel treatment in patients with st-segment elevation myocardial infarction undergoing primary percutaneous coronary intervention - . pubmed [internet]. available at: https://pubmed.ncbi.nlm.nih.gov/21719452/

[153] . Dörler j, edlinger m, alber hf, altenberger j, benzer w, grimm g, et al. clopidogrel pre-treatment is associated with reduced in-hospital mortality in primary percutaneous coronary intervention for acute st-elevation myocardial infarction. eurheart j. dec 2011;32(23):2954-61.

[154] . Zeymer u, arntz hr, mark b, fichtlscherer s, werner g, schöller r, et al. efficacy and safety of a high loading dose of clopidogrel administered prehospitally to improve primary percutaneous coronary intervention in acute myocardial infarction: the randomized cipami trial. clin res cardiol. 1 Apr 2012;101(4):305-12.

[155] . Wiviott sd, braunwald e, mccabe ch, montalescot g, ruzyllo w, gottlieb s, et al. prasugrel versus clopidogrel in patients with acute coronary syndromes. n engl jmed. 15 nov 2007;357(20):2001-15.

[156] . Wallentin l, becker rc, budaj a, cannon cp, emanuelsson h, held c, et al. ticagrelor versus clopidogrel in patients with acute coronary syndromes. n engl j med. 10 sept 2009;361(11):1045-57.

[157] . Mehta sr, tanguay jf, eikelboom jw, jolly ss, joyner cd, granger cb, et al. doubledose versus standard-dose clopidogrel and high-dose versus low-dose aspirin in individuals undergoing percutaneous coronary intervention for acute coronary syndromes (current-oasis 7): a randomised factorial trial. lancet. 9 oct 2010;376(9748): 1233-43.

[158] . Steg pg, bhatt dl, hamm cw, stone gw, gibson cm, mahaffey kw, et al. effect of cangrelor on periprocedural outcomes in percutaneous coronary interventions: a pooled analysis of patient-level data. lancet. 14 Dec 2013;382(9909):1981-92.

[159] . Valgimigli m, ariotti s, costa f. duration of dual antiplatelet therapy after drugeluting stent implantation: will we ever reach a consensus? eur heart j. 21 may 2015;36(20):1219-22.

[160] . Costa f, tijssen jg, ariotti s, giatti s, moscarella e, guastaroba p, et al. incremental value of the crusade, acuity, and has-bled risk scores for the prediction of hemorrhagic events after coronary stent implantation in patients undergoing long or short duration of dual antiplatelet therapy. j am heart assoc. 7 dec 2015;4(12):e002524.

[161] . Chen zm, pan hc, chen yp, peto r, collins r, jiang lx, et al. early intravenous then oral metoprolol in 45,852 patients with acute myocardial infarction: randomised placebo-controlled trial. lancet. 5 nov 2005;366(9497):1622-32.

[162] . Pfisterer m, cox jl, granger cb, brener sj, naylor cd, califf rm, et al. atenolol use and clinical outcomes after thrombolysis for acute myocardial infarction: the gusto-i experience. global utilization of streptokinase and tpa (alteplase) for occluded coronary arteries. j am coll cardiol. sept 1998;32(3):634-40.

[163] . Chatterjee s, chaudhuri d, vedanthan r, fuster v, ibanez b, bangalore s, et al. early intravenous beta-blockers in patients with acute coronary syndrome--a meta-analysis of randomized trials. int j cardiol. 30 Sep 2013;168(2):915-21.

[164] . Ibanez b, macaya c, sánchez-brunete v, pizarro g, fernández-friera l, mateos a, etal. effect of early metoprolol on infarct size in st-segment-elevation myocardial infarction patients undergoing primary percutaneous coronary intervention: the effect of metoprolol in cardioprotection during an acute myocardial infarction (metocard-cnic) trial. circulation. 1 Oct 2013;128(14):1495-503.

[165] . Pizarro g, fernández-friera l, fuster v, fernández-jiménez r, garcía-ruiz jm, garcía-álvarez a, et al. long-term benefit of early pre-reperfusion metoprolol administration in patients with acute myocardial infarction: results from the metocard-cnic trial (effect of metoprolol in cardioprotection during an acute myocardial infarction). j am coll cardiol. 10 june 2014;63(22):2356-62.

[166] . García-prieto j, villena-gutiérrez r, gómez m, bernardo e, pun-garcía a, garcía-lunar i, et al. neutrophil stunning by metoprolol reduces infarct size. nat commun. 18 Apr 2017;8:14780.

[167] . Hawkins cm, richardson dw, vokonas ps. effect of propranolol in reducing mortality in older myocardial infarction patients. the beta-blocker heart attack trial experience. circulation. june 1983;67(6 pt 2):i94-97.

[168] . Olsson g, rehnqvist n, sjögren a, erhardt l, lundman t. long-term treatment with metoprolol after myocardial infarction: effect on 3 year mortality and morbidity. j am coll cardiol. june 1985;5(6):1428-37.

[169] . Norwegian multicenter study group. timolol-induced reduction in mortality and reinfarction in patients surviving acute myocardial infarction. n engl j med. 2 Apr 1981;304(14):801-7.

[170] . Freemantle n, cleland j, young p, mason j, harrison j. beta blockade after myocardial infarction: systematic review and meta regression analysis. bmj. 26 June 1999;318(7200):1730-7.

[171] . Goldberger jj, bonow ro, cuffe m, liu l, rosenberg y, shah pk, et al. effect of betablocker dose on survival after acute myocardial infarction. j am coll cardiol. 29 Sep 2015;66(13):1431-41.

[172] . Randomised trial of intravenous atenolol among 16 027 cases of suspected acute myocardial infarction: isis-1. first international study of infarct survival collaborative group. lancet. 12 jul 1986;2(8498):57-66.

[173] . Coats aj. capricorn: a story of alpha allocation and beta-blockers in left ventricular dysfunction post-mi. int j cardiol. apr 2001;78(2):109-13.

[174] . Dondo tb, hall m, west rm, jernberg t, lindahl b, bueno h, et al. ß-blockers and mortality after acute myocardial infarction in patients without heart failure or ventricular dysfunction. j am coll cardiol. 6 June 2017;69(22):2710-20.

[175] . Bugiardini r, cenko e, ricci b, vasiljevic z, dorobantu m, kedev s, et al. comparison of early versus delayed oral ß blockers in acute coronary syndromes and effect on outcomes. am j cardiol. 1 march 2016;117(5):760-7.

[176] . Indications for ace inhibitors in the early treatment of acute myocardial infarction: systematic overview of individual data from 100,000 patients in randomized trials. ace inhibitor myocardial infarction collaborative group. circulation. 9 June 1998;97(22):2202-12.
[177] . Saha sa, molnar j, arora rr. tissue ace inhibitors for secondary prevention of cardiovascular disease in patients with preserved left ventricular function: a pooled metaanalysis of randomized placebo-controlled trials. j cardiovasc pharmacol ther. sept 2007;12(3):192-204.
[178] . Hikosaka m, yuasa f, yuyama r, motohiro m, mimura j, kawamura a, et al. effect of angiotensin-converting enzyme inhibitor on cardiopulmonary baroreflex sensitivity in patients with acute myocardial infarction. am j cardiol. 1 dec 2000;86(11):1241-4, a6.
[179] . Fox km, european trial on reduction of cardiac events with perindopril in stable coronary artery disease investigators. efficacy of perindopril in reduction of cardiovascular events among patients with stable coronary artery disease: randomised, double-blind, placebo-controlled, multicentre trial (the europa study). lancet. 6 sept 2003;362(9386):782-8.
[180] . Heart outcomes prevention evaluation study investigators, yusuf s, sleight p, pogue j, bosch j, davies r, et al. effects of an angiotensin-converting-enzyme inhibitor, ramipril, on cardiovascular events in high-risk patients. n engl j med. 20 janv 2000;342(3):145-53.
[181] . Pfeffer ma, mcmurray jjv, velazquez ej, rouleau jl, k0ber l, maggioni ap, et al. valsartan, captopril, or both in myocardial infarction complicated by heart failure, left ventricular dysfunction, or both. n engl j med. 13 nov 2003;349(20):1893-906.
[182] . Ridker pm, danielson e, fonseca fah, genest j, gotto am, kastelein jjp, et al. rosuvastatin to prevent vascular events in men and women with elevated c-reactive protein. n engl j med. 20 nov 2008;359(21):2195-207.
[183] . Randomised trial of cholesterol lowering in 4444 patients with coronary heart disease: the scandinavian simvastatin survival study (4s). lancet. 19 nov 1994;344(8934):1383-9.
[184] . Sacks fm, pfeffer ma, moye la, rouleau jl, rutherford jd, cole tg, et al. the effect of pravastatin on coronary events after myocardial infarction in patients with average cholesterol levels. cholesterol and recurrent events trial investigators. n engl j med. 3 oct 1996;335(14):1001-9.
[185] . Baigent c, keech a, kearney pm, blackwell l, buck g, pollicino c, et al. efficacy and safety of cholesterol-lowering treatment: prospective meta-analysis of data from 90,056 participants in 14 randomised trials of statins. lancet. 8 oct 2005;366(9493):1267-78.
[186] . Serban mc, colantonio ld, manthripragada ad, monda kl, bittner va, banach m, et al. statin intolerance and risk of coronary heart events and all-cause mortality following myocardial infarction. j am coll cardiol. 21 march 2017;69(11):1386-95.
[187] . Boekholdt sm, hovingh gk, mora s, arsenault bj, amarenco p, pedersen tr, et al. very low levels of atherogenic lipoproteins and the risk for cardiovascular events: a metaanalysis of statin trials. j am coll cardiol. 5 august 2014;64(5):485-94.
[188] . Cholesterol treatment trialists' (ctt) collaboration, fulcher j, o'connell r, voysey m, emberson j, blackwell l, et al. efficacy and safety of ldl-lowering therapy among men and women: meta-analysis of individual data from 174,000 participants in 27 randomised

trials. lancet. 11 Apr 2015;385(9976):1397-405.
[189] . Cohen jc, boerwinkle e, mosley th, hobbs hh. sequence variations in pcsk9, low ldl, and protection against coronary heart disease. n engl j med. 23 march 2006;354(12):1264-72.
[190] . Sabatine ms, giugliano rp, keech ac, honarpour n, wiviott sd, murphy sa, et al. evolocumab and clinical outcomes in patients with cardiovascular disease. n engl jmed. May 4, 2017;376(18):1713-22.
[191] . Ray kk, landmesser u, leiter la, kallend d, dufour r, karakas m, et al. inclisiran in patients at high cardiovascular risk with elevated ldl cholesterol. new england journal of medicine. 13 Apr 2017;376(15):1430-40.
[192] . Hypolipidemics: how to prescribe? [internet]. available on: https://www.cardio-online.fr/actualites/a-la-une/comment-prescrire-traitement- lipid-lowering drugs
[193] . Schwartz gg, steg pg, szarek m, bhatt dl, bittner va, diaz r, et al. alirocumab and cardiovascular outcomes after acute coronary syndrome. n engl j med. 29 nov 2018;379(22):2097-107.
[194] . Yusuf s, held p, furberg c. update of effects of calcium antagonists in myocardial infarction or angina in light of the second danish verapamil infarction trial (davit- ii) and other recent studies. am j cardiol. 1 june 1991;67(15):1295-7.
[195] . Held ph, yusuf s, furberg cd. calcium channel blockers in acute myocardial infarction and unstable angina: an overview. bmj. 11 nov 1989;299(6709): 1187-92.
[196] . Effect of verapamil on mortality and major events after acute myocardial infarction (the danish verapamil infarction trial ii--davit ii). am j cardiol. 1 oct 1990;66(10):779-85.
[197] . Furberg cd, psaty bm, meyer jv. nifedipine. dose-related increase in mortality in patients with coronary heart disease. circulation. 1 sept 1995;92(5):1326-31.
[198] . Poole-wilson pa, lubsen j, kirwan ba, van dalen fj, wagener g, danchin n, et al. effect of long-acting nifedipine on mortality and cardiovascular morbidity in patients with stable angina requiring treatment (action trial): randomised controlled trial. lancet. 4 sept 2004;364(9437):849-57.
[199] . Isis-4: a randomised factorial trial assessing early oral captopril, oral mononitrate, and intravenous magnesium sulphate in 58,050 patients with suspected acute myocardial infarction. isis-4 (fourth international study of infarct survival) collaborative group. lancet. March 18, 1995;345(8951):669-85.
[200] . Pitt b, remme w, zannad f, neaton j, martinez f, roniker b, et al. eplerenone, a selective aldosterone blocker, in patients with left ventricular dysfunction after myocardial infarction. n engl j med. 3 apr 2003;348(14):1309-21.
[201] . Pitt b, zannad f, remme wj, cody r, castaigne a, perez a, et al. the effect of spironolactone on morbidity and mortality in patients with severe heart failure. randomized aldactone evaluation study investigators. n engl j med. 2 sept 1999;341(10):709-17.
[202] . Montalescot g, pitt b, lopez de sa e, hamm cw, flather m, verheugt f, et al. early eplerenone treatment in patients with acute st-elevation myocardial infarction without heart failure: the randomized double-blind reminder study. eur heart j. 7 sept 2014;35(34):2295-302.
[203] . Beygui f, cayla g, roule v, roubille f, delarche n, silvain j, et al. early aldosterone blockade in acute myocardial infarction: the albatross randomized clinical trial. j am coll

cardiol. 26 Apr 2016;67(16):1917-27.
[204] . Ross r. atherosclerosis--an inflammatory disease. n engl j med. 14 jan 1999;340(2):115-26.
[205] . Ridker pm, everett bm, thuren t, macfadyen jg, chang wh, ballantyne c, et al. antiinflammatory therapy with canakinumab for atherosclerotic disease. n engl j med. 21 Sep 2017;377(12):1119-31.
[206] . Abbate a, van tassell bw, biondi-zoccai g, kontos mc, grizzard jd, spillmandw, et al. effects of interleukin-1 blockade with anakinra on adverse cardiac remodeling and heart failure after acute myocardial infarction [from the virginia commonwealth university-anakinra remodeling trial (2) (vcu-art2) pilot study]. am j cardiol. 15 may 2013;111(10):1394-400.
[207] . Abbate a, kontos mc, grizzard jd, biondi-zoccai ggl, van tassell bw, robati r, et al. interleukin-1 blockade with anakinra to prevent adverse cardiac remodeling after acute myocardial infarction (virginia commonwealth university anakinra remodeling trial [vcu-art] pilot study). am j cardiol. 15 may 2010;105(10):1371- 1377.e1.
[208] . Rymer ja, newby lk. failure to launch: targeting inflammation in acute coronary syndromes. jacc basic transl sci. aoùt 2017;2(4):484-97.
[209] . Reinstadler sj, stiermaier t, fuernau g, de waha s, desch s, metzler b, et al. the challenges and impact of microvascular injury in st-elevation myocardial infarction. expert rev cardiovasc ther. 2016;14(4):431-43.
[210] . Papapostolou s, andrianopoulos n, duffy sj, brennan al, ajani ae, clark dj, et al. long-term clinical outcomes of transient and persistent no-reflow following percutaneous coronary intervention (pci): a multicentre australian registry. eurointervention. 20 June 2018;14(2):185-93.
[211] . Feher a, chen sy, bagi z, arora v. prevention and treatment of no-reflow phenomenon by targeting the coronary microcirculation. rev cardiovasc med. 2014;15(1):38-51.
[212] . Fischell ta, fischell dr, avezum a, john ms, holmes d, foster m, et al. initial clinical results using intracardiac electrogram monitoring to detect and alert patients during coronary plaque rupture and ischemia. j am coll cardiol. 28 sept2010;56(14):1089-98.
[213] . Us Food & drug administration. fda approval letter angelmed guardian system (p150009). fda.gov https://www.accessdata.fda.gov/cdrh_docs/pdf15/ p150009a.pdf (2018).
[214] . Schmidt mr, rasmussen me, b0tker he. remote ischemic conditioning for patients with stemi. j cardiovasc pharmacol ther. jul 2017;22(4):302-9.
[215] . Stone gw, selker hp, thiele h, patel mr, udelson je, ohman em, et al. relationship between infarct size and outcomes following primary pci: patient- level analysis from 10 randomized trials. j am coll cardiol. 12 Apr 2016;67(14):1674-83.
[216] . Ibáñez b, heusch g, ovize m, van de werf f. evolving therapies for myocardial ischemia/reperfusion injury. j am coll cardiol. 14 Apr 2015;65(14):1454-71.
[217] . Niccoli g, scalone g, lerman a, crea f. coronary microvascular obstruction in acute myocardial infarction. eur heart j. 1 Apr 2016;37(13):1024-33.
[218] . Hausenloy dj, botker he, engstrom t, erlinge d, heusch g, ibanez b, et al. targeting reperfusion injury in patients with st-segment elevation myocardial infarction: trials and tribulations. eur heart j. 1 Apr 2017;38(13):935-41.
[219] . Schmitt j, duray g, gersh bj, hohnloser sh. atrial fibrillation in acute myocardial

infarction: a systematic review of the incidence, clinical features and prognostic implications. eur heart j. may 2009;30(9):1038-45.

[220] . Batra g, svennblad b, held c, jernberg t, johanson p, wallentin l, et al. all types of atrial fibrillation in the setting of myocardial infarction are associated with impaired outcome. heart. 15 June 2016;102(12):926-33.

[221] . Kirchhof p, benussi s, kotecha d, ahlsson a, atar d, casadei b, et al. 2016 esc guidelines for the management of atrial fibrillation developed in collaboration with eacts. eur j cardiothorac surg. nov 2016;50(5):e1-88.

[222] . Nilsson kr, al-khatib sm, zhou y, pieper k, white hd, maggioni ap, et al. atrial fibrillation management strategies and early mortality after myocardial infarction: results from the valsartan in acute myocardial infarction (valiant) trial. heart. june 2010;96(11):838-42.

[223] . Gorenek b, blomström lundqvist c, brugada terradellas j, camm aj, hindricks g, huber k, et al. cardiac arrhythmias in acute coronary syndromes: position paper from the joint ehra, acca, and eapci task force. europace. nov 2014;16(11):1655-73.

[224] . Jabre p, jouven x, adnet f, thabut g, bielinski sj, weston sa, et al. atrial fibrillation and death after myocardial infarction: a community study. circulation. 17 May 2011;123(19):2094-100.

[225] . Siu cw, jim mh, ho hh, miu r, lee swl, lau cp, et al. transient atrial fibrillation complicating acute inferior myocardial infarction: implications for future risk of ischemic stroke. chest. jul 2007;132(1):44-9.

[226] . Demidova mm, smith jg, höijer cj, holmqvist f, erlinge d, platonov pg. prognostic impact of early ventricular fibrillation in patients with st-elevation myocardial infarction treated with primary pci. eur heart j acute cardiovasc care. dec 2012;1(4):302-11.

[227] . Mehta rh, starr az, lopes rd, hochman js, widimsky p, pieper ks, et al. incidence of and outcomes associated with ventricular tachycardia or fibrillation in patients undergoing primary percutaneous coronary intervention. jama. 6 May 2009;301(17):1779-89.

[228] . Piccini jp, schulte pj, pieper ks, mehta rh, white hd, van de werf f, et al. antiarrhythmic drug therapy for sustained ventricular arrhythmias complicating acute myocardial infarction. crit care med. jan 2011;39(1):78-83.

[229] . Dumas f, cariou a, manzo-silberman s, grimaldi d, vivien b, rosencher j, et al. immediate percutaneous coronary intervention is associated with better survival after out-of-hospital cardiac arrest: insights from the procat (parisian region out of hospital cardiac arrest) registry. circ cardiovasc interv. 1 June 2010;3(3):200-7.

[230] . V R, b i, jp o, g p, n van r, a m, et al. early intravenous beta-blockers in patients with st-segment elevation myocardial infarction before primary percutaneous coronary intervention. journal of the american college of cardiology [internet]. 14 June 2016 ;67(23). available from: https://pubmed.ncbi.nlm.nih.gov/27050189/

[231] . Piccini jp, hranitzky pm, kilaru r, rouleau jl, white hd, aylward pe, et al. relation of mortality to failure to prescribe beta blockers acutely in patients with sustained ventricular tachycardia and ventricular fibrillation following acute myocardial infarction (from the valsartan in acute myocardial infarction trial [valiant] registry). am j cardiol. 1 dec 2008;102(11):1427-32.

[232] . Zafari am, zarter sk, heggen v, wilson p, taylor ra, reddy k, et al. a program encouraging early defibrillation results in improved in-hospital resuscitation efficacy. j am coll cardiol. 18 august 2004;44(4):846-52.

[233] . Wolfe cl, nibley c, bhandari a, chatterjee k, scheinman m. polymorphous ventricular tachycardia associated with acute myocardial infarction. circulation. oct 1991;84(4):1543-51.

[234] . Liang jj, fender ea, cha ym, lennon rj, prasad a, barsness gw. long-term outcomes in survivors of early ventricular arrhythmias after acute st-elevation and non-st- elevation myocardial infarction treated with percutaneous coronary intervention. am j cardiol. 1 march 2016;117(5):709-13.

[235] . Masuda m, nakatani d, hikoso s, suna s, usami m, matsumoto s, et al. clinical impact of ventricular tachycardia and/or fibrillation during the acute phase of acute myocardial infarction on in-hospital and 5-year mortality rates in the percutaneous coronary intervention era. circ j. 24 june 2016;80(7):1539-47.

[236] . Priori sg, blomström-lundqvist c, mazzanti a, blom n, borggrefe m, camm j, et al. 2015 esc guidelines for the management of patients with ventricular arrhythmias and the prevention of sudden cardiac death: the task force for the management of patients with ventricular arrhythmias and the prevention of sudden cardiac death of the european society of cardiology (esc). endorsed by: association for european paediatric and congenital cardiology (aepc). eur heart j. 1 nov 2015;36(41):2793-867.

[237] . Zimetbaum pj, josephson me. use of the electrocardiogram in acute myocardial infarction. new england journal of medicine. 6 march 2003;348(10):933-40.

[238] . Meine tj, al-khatib sm, alexander jh, granger cb, white hd, kilaru r, et al. incidence, predictors, and outcomes of high-degree atrioventricular block complicating acute myocardial infarction treated with thrombolytic therapy. am heart j. apr 2005;149(4):670-4.

[239] . Gang ujo, hvelplund a, pedersen s, iversen a, j0ns c, abildstom sz, et al. high-degree atrioventricular block complicating st-segment elevation myocardial infarction in the era of primary percutaneous coronary intervention. europace. nov 2012;14(11):1639-45.

[240] . Feigl d, ashkenazy j, kishon y. early and late atrioventricular block in acute inferior myocardial infarction. j am coll cardiol. juill 1984;4(1):35-8.

[241] . Bertolet bd, mcmurtrie eb, hill ja, belardinelli l. theophylline for the treatment of atrioventricular block after myocardial infarction. ann intern med. 1 oct 1995;123(7):509-11.

[242] . Kim Kh, jeong mh, ahn y, kim yj, cho mc, kim w, et al. diffrential clinical implications of high-degree atrioventricular block complicating st-segment elevation myocardial infarction according to the location of infarction in the era of primary percutaneous coronary intervention. korean circ j. may 2016;46(3):315-23.

[243] . Mccarthy cp, vaduganathan m, mccarthy kj, januzzi jl, bhatt dl, mcevoy jw. left ventricular thrombus after acute myocardial infarction: screening, prevention, and treatment. jama cardiol. 1 jul 2018;3(7):642-9.

[244] . Pöss j, desch s, eitel c, de waha s, thiele h, eitel i. left ventricular thrombus formation after st-segment-elevation myocardial infarction: insights from a cardiac magnetic resonance multicenter study. circ cardiovasc imaging. oct 2015;8(10):e003417.

[245] . Cambronero-cortinas e, bonanad c, monmeneu jv, lopez-lereu mp, gavara j, de dios e, et al. incidence, outcomes, and predictors of ventricular thrombus after reperfused st-segment-elevation myocardial infarction by using sequential cardiac mr imaging. radiology. august 2017;284(2):372-80.

[246] . Weinsaft jw, kim j, medicherla cb, ma cl, codella ncf, kukar n, et al. echocardiographic algorithm for post-myocardial infarction lv thrombus: a gatekeeper for thrombus evaluation by delayed enhancement cmr. jacc cardiovasc imaging. may 2016;9(5):505-15.

[247] . Bulluck h, dharmakumar r, arai ae, berry c, hausenloy dj. cardiovascular magnetic resonance in acute st-segment-elevation myocardial infarction: recent advances, controversies, and future directions. circulation. May 1, 2018;137(18):1949-64.

[248] . Gellen b, biere l, logeart d, lairez o, vicaut e, furber a, et al. timing of cardiac magnetic resonance imaging impacts on the detection rate of left ventricular thrombus after myocardial infarction. jacc cardiovasc imaging. nov 2017;10(11):1404-5.

[249] . Meurin p, brandao carreira v, dumaine r, shqueir a, milleron o, safar b, et al. incidence, diagnostic methods, and evolution of left ventricular thrombus in patients with anterior myocardial infarction and low left ventricular ejection fraction: a prospective multicenter study. am heart j. aoüt 2015;170(2):256-62.

[250] . Zhang z, si d, zhang q, jin l, zheng h, qu m, et al. prophylactic rivaroxaban therapy for left ventricular thrombus after anterior st-segment elevation myocardial infarction. jacc cardiovasc interv. 25 Apr 2022;15(8):861-72.

[251] . French jk, hellkamp as, armstrong pw, cohen e, kleiman ns, o'connor cm, et al. mechanical complications after percutaneous coronary intervention in st-elevation myocardial infarction (from apex-ami). am j cardiol. 1 jan 2010;105(1):59-63.

[252] . Mclaughlin a, mcgiffin d, winearls j, tesar p, cole c, vallely m, et al. veno-arterial ecmo in the setting of post-infarct ventricular septal defect: a bridge to surgical repair. heart lung circ. nov 2016;25(11):1063-6.

[253] . Durko ap, budde rpj, geleijnse ml, kappetein ap. recognition, assessment and management of the mechanical complications of acute myocardial infarction. heart. jul 2018;104(14):1216-23.

[254] . Kolte d, khera s, aronow ws, mujib m, palaniswamy c, sule s, et al. trends in incidence, management, and outcomes of cardiogenic shock complicating st- elevation myocardial infarction in the united states. j am heart assoc. 13 jan 2014;3(1):e000590.

[255] . Auffret v, leurent g, gilard m, hacot jp, filippi e, delaunay r, et al. incidence, timing, predictors and impact of acute heart failure complicating st-segment elevation myocardial infarction in patients treated by primary percutaneous coronary intervention. int j cardiol. 15 oct 2016;221:433-42.

[256] . Thiele h, ohman em, desch s, eitel i, de waha s. management of cardiogenic shock. eur heart j. 21 may 2015;36(20):1223-30

[257] . Bellemain-appaix a, collet jp, montalescot g. acute coronary syndromes. emc (elsevier masson sas, paris), cardiology, 11-030-d-10, 2010.

[258] . Collège des enseignants de cardiologie et maladies vasculaires, cardiovascular risk factors and prevention, French-speaking virtual medical university, 2011-2012.

[259] . **Boersma e et al.** early thrombolytic treatment in acute myocardial infarction: reappraisal of the golden hour. lancet 1996; 348:771-775.

[260] . **Emip.**the european myocardial infarction project group. prehospital thrombolytic therapy in patients with suspected acute myocardial infarction.n.engl.j.med 1993; 329: 383-9

[261] . **Miti.** the myocardial infarction triage and intervention project group. jama 1993; 270:1211-1216.

[262] . **-Late study group:** late assessment of thrombolytic efficacy (late) study with alteplase 6-24 hours after onset of acute myocardial infarction. lancet 1993;342:759-766

[263] . **-Gusto angiographic investigators.** the effects of tissue plasminogen avtivator , streptokinase ,or both coronary -artery patency , ventricular function ,and survival after acute myocardial .n.engl .j med 1993 ; 329 : 1615 - 22

[264] . **-Gruppo italiano per lo studio della sopravivenza nell'infarto miocardico . (gissi-2):** a factorial randomised trial of alteplase verus streptokinase and heparin versus no heparin among 12 490 patients with acute myocardial infarction.lancet 1990; 336 : 65 - 71

[265] . **-Isis-3 (third international study of infarct survival) collaborative study group .**randomised comparison for streptokinase vs tissue plasminogen activator vs anistreplase and of aspirin plus heparin vs aspirin alone among 41 299 cases of suspected acute myocardial infarction. lancet 1992 ; 339 : 75370

[266] . **-Gusto investigators** an international randomised trial comparing thrombolytic strategies for acute myocardial infarction. new england journal of medicine 1993; 329 : 673 - 82

[267] . **-Gusto iib.** the global use of strategies to open occluded coronary arteries in acute coronary syndromes (gusto iib) angioplasty substudy investigators. n. engl. j med 1997; 23: 1621-1628.

[268] . **-Gusto iii.** the global use of strategies to open occluded coronary arteries (gusto iii) investigators. a comparison of reteplase with alteplase for acute myocardial infarction. n. engl. j med 1997; 337: 1118-1123.

[269] . **-Timi 10 a.** circulation 1999; 95:351-356.

[270] . **-Timi 10 b.**circulation 1998;98:**2805-2814**.

[271] . **-Assent 1.** assessment of the safety and efficacy of a new thrombolytic agent. am heart j 1999; 137:786-791.

[272] . **-Assent-2.** single bolus tenecteplase compared with front-loaded alteplase in acute myocardial infarction: the assent-2 double blind randomised trial . lancet 1999; 354:71622.

[273] . Thiele h, jobs a, ouweneel dm, henriques jps, seyfarth m, desch s, et al. percutaneous short-term active mechanical support devices in cardiogenic shock: a systematic review and collaborative meta-analysis of randomized trials. eur heart j. 14 Dec 2017;38(47):3523-31.

[274] . Pinto, d. s. et al. benefit of transferring st-segment-elevation myocardial infarction patients for percutaneous coronary intervention compared with administration of onsite fibrinolytic declines as delays increase. circulation [internet]. 12 June 2011;124(23). available at: https://pubmed.ncbi.nlm.nih.gov/22064592/

[275] . Bonnefoy e, steg pg, boutitie f, dubien py, lapostolle f, roncalli j, et al. comparison of primary angioplasty and pre-hospital fibrinolysis in acute myocardial infarction (captim) trial: a 5-year follow-up. eur heart j. juill 2009;30(13):1598-606.

[276] . Armstrong pw, gershlick ah, goldstein p, wilcox r, danays t, lambert y, et al. fibrinolysis or primary pci in st-segment elevation myocardial infarction. n engl j med. 11 Apr 2013;368(15):1379-87.

[277] . Khan m, hashim m, mustafa h, et al. (july 23, 2020) global epidemiology of ischemic heart disease: results from the global burden of diseasesstudy. cureus 12(7): e9349. doi 10.7759/cureus.9349

[278] . Effects of clopidogrel in addition to aspirin in patients with acute coronary syndromes without st-segment elevation. n engl j med. 2001;345(7):494- 502.

[279] . Wiviott sd, braunwald e, mccabe ch, montalescot g, ruzyllo w, gottlieb s, et al. prasugrel versus clopidogrel in patients with acute coronary syndromes. new england journal of medicine. 15 nov 2007;357(20):2001- 15.

[280] . Wallentin l, becker rc, budaj a, cannon cp, emanuelsson h, held c, et al. ticagrelor versus clopidogrel in patients with acute coronary syndromes. n engl j med. 10 sept 2009;361(11):1045- 57.

[281] . Brandler e, paladino l, sinert r. does the early administration of beta-blockers improve the in-hospital mortality rate of patients admitted with acute coronary syndrome? acad emerg med. jan 2010;17(1):1- 10.

[282] . Dirkali a, van der ploeg t, nangrahary m, cornel jh, umans vawm. the impact of admission plasma glucose on long-term mortality after stemi and nstemi myocardial infarction. international journal of cardiology. 1 oct 2007;121(2):215- 7.

[283] . Ling y, li x, gao x. intensive versus conventional glucose control in critically ill atients: a meta-analysis of randomized controlled trials. european journal of internal medicine. sept 2012;23(6):564- 74.

[284] . Grines c.l., browne k.f., marco j. et al. a comparison of immediate angioplasty with thrombolytic therapy for acute myocardial infarction. n engl j med 1993; 338:673679.

[285] . Weaver d.w., simes j., betriu et al. comparison of primary coronary angioplasty and intravenous thrombolytic theray for acute myocardial infarction. a quantitativereview. jama 1997; 278: 2093-2098.

[286] . Wallentin l.c. reducing time to treatment in acute myocardial infarction. eur j emerg med 2000; 7: 217-227.

[287] . Strategic reperfusion early after myoocardial infarction (stream) study am_hrt_journal_2010_v160_n1

[288] . Boersma e., maas a.c., deckers j.w., simoons m.l. early thrombolytic treatment in acute myocardial infarction: reappraisal of the golden hour. lancet 1996; 348: 771-775.

[289] . Linderer t., schroder r., arntz r. et al. prehospital thrombolysis: beneficial effects of very early treatment on infarct size and left ventricular function. j am coll cardiol1993 ; 126: 832-839.

[290] . Morrisson l.j., verbeek p.r., mc donald a.c. et al. mortality and pre hospital thrombolysis for acute myocardial infarction. a meta-analysis. jama 2000 ; 283 : 20 :2686-2892.

[291] . Lapandry c., laperce t., lambert y. et al. prehospital management of st+ acute coronary syndromes in ile-de-france. the e- must register. arch mal coeur vaiss2005; 98: 1137-1142.

[292] . Ftt collaborative group indications for fibrinolytic therapy in suspected acute myocardial infarction: collaborative overview of early mortality and major morbidity results from all randomised trials of more than 1000 patients. fibrinolytic therapy trialists' (ftt) collaborative group. lancet 1994; 343: 311-322.

[293] . Hochma j.s., sleeper l.a., webb j.g. et al. early revascularization in acute myocardial infarction complicated by cardiogenic shock. n engl j med 1999; 344: 625634.

[294] . **-Letouzet jp, genet a, amoretti r.** white paper on the management of cardiovascular disease in france. cardiologie ***2000*** 1996: 13-24.

[295] . Haouchine n, mohand-oussaid s, benyoussef n, boukhroufa f, nibouche d, redjimi m. study of a series of acute myocardial infarctions during 2004.

[296] . Société algérienne de cardiologie. st elevation algeria myocardial infarction (stami).

[297] . Boussouf k, zaidi z, kaddour f, djelaoudji a, benkobbi s, bayadi n, et al. clinical epidemiology of acute myocardial infarction in setif, algeria: finding from the setif-ami registry. health sci j [internet]. 2019 [cited 25 March 2021];13(1). available from: http://www.hsj.gr/medicine/clinical-epidemiology- of-acute-myocardial-infarction-in-setif-algeria-finding-from-the-setifami- registry.php?aid=24161.

[298] . Laraba n. prevalence, predictive factors and impact of polyvascular disease in patients with acute coronary syndrome [thesis]. alger: université d'alger 1; 2016.

[299] . Milouchi s, ajmi h, sghaier a, khorchani a, ferjani s. management of acute coronary syndrome with st segment elevation in medenine regional hospital. a propos de 150 cas. cardiologie tunisienne.2eme trimestre 2017; 13(02) :93-100.

[300] . N'guetta r, yao h, ekou a, n'cho-mottoh mp, angoran i, tano m, et al. prevalence and characteristics of acute coronary syndromes in a sub-Saharan african population. annals of cardiology and angiology. apr 2016;65(2):59-63.

[301] . Gouda m, ammar as, naguib ta, ateya aa, ibrahim e-sa. incidence and criteria of acute coronary syndrome among the population of north sinai governorate. j heart res. nov 2018;1:14-7.

[302] . **Richard jl.** le projet monica. un projet oms de recherche cardiovasculaire. rev epidemiol santé publiquee 1988 ; 36 :325-34.

[303] . Mcmanus dd, gore j, yarzebski j, spencer f, lessard d, goldberg rj. recent trends in the incidence, treatment, and outcomes of patients with stemi and nstemi. the american journal of medicine. jan 2011;124(1):40-7.

[304] . Belle l, cayla g, cottin y, coste p, khalife k, labèque jn, et al. french registry on acute st-elevation and non-st-elevation myocardial infarction 2015 (fast-mi 2015). design and baseline data. arch cardiovasc dis. july 2017;110(6-7):366-78.

[305] . Addad f, mahdhaoui a, gouider j, boughzela e, kamoun s, boujnah mr, et al. management of patients with acute st-elevation myocardial infarction: results of the fast-mi tunisia registry. plos one. 2019;14(2):e0207979.

[306] . Akoudad h, el khorb n, sekkali n, mechrafi a, zakari n, ouaha l, et al. l'infarctus du myocarde au maroc: les données du registre fes-ami. annales de cardiologie et d'angéiologie. 1 Dec 2015;64(6):434-8.

[307] . Moustaghfir a, haddak m, mechmeche r. management of acute coronarysyndromes in maghreb countries: the access (acute coronary events - a multinational survey of current management strategies) registry. archives of cardiovascular diseases. 1 nov 2012;105(11):566-77.

[308] . Steyn k, sliwa k, hawken s, commerford p, onen c, damasceno a, et al. risk factors associated with myocardial infarction in africa. the interheart africa study. circulation. 6 Dec 2005;112(23):3554-61.

[309] . Al Thani h, el-menyar a, alhabib kf, al-motarreb a, hersi a, alfaleh h, et al. polyvascular disease in patients presenting with acute coronary syndrome: its predictors and outcomes. the scientific world journal. 2012;2012:1-7.

[310] . Mukherjee d, eagle ka, kline-rogers e, feldman lj, juliard j-m, agnelli g, et al. impact of prior peripheral arterial disease and stroke on outcomes of acute coronary syndromes and effect of evidence-based therapies (from the global registry of acute coronary events). americanjournal of cardiology. 1 july 2007;100(1):1-6.

[311] . Bhatt dl, peterson ed, harrington ra, ou f-s, cannon cp, gibson cm, et al. prior polyvascular disease: risk factor for adverse ischaemic outcomes in acute coronary syndromes. european heart journal. 1 May 2009;30(10):1195-202.

[312] . -Johansson s, bergstrand r, ulvenstam g, vedin a et al. sex differences in preinfarction characteristics and long- term survival among patients with myocardial infarction. am j epidemiol 1984; 119: 610-623.

[313] . Belgacem abdelkrim mohamed, reperfusion medicamenteuse dans l'infarctus du myocarde aigu : évaluation clinique et benefice therapeutique, thesis pour le diplôme de docteur en sciences medicales, 2011.

[314] . -Juliard j .m .,himbert d .,aubry p ., benamer h ., karrilon g.j ., boccara . feldman l.j ., streg p.g, prise en charge orientée vers la reperfusion en phase aigue d'infarctus du myocarde .résultats sur une cohorte de 700 patients consécutifs .archives des maladies du creur et des vaisseaux 1997 ;90 (3) : 337 -43 .

[315] . -De benedetti e, urban p, burgan s, dorsaz pa, chatelain p, gaspoz j-m, chevrolet j-c, unger p-f. thrombolysis in acute myocardial infarction in routine clinical practice. schweiz med wochenschr 1997; 127: 1285-1290.

[316] . -Bougatef s, ben romdhane h, kafsi n, belhani a, haouala h, slimane ml, boujnah r, drissa h , achour n, gueddiche m. delay in treatment and access to care for myocardial infarction in hospitals: results of a multicentre study. journal de la société tunisienne des sciences médicales, 2005, 83 (supp 5) :19-23.

[317] . -Gabriel steg p, laperche t, golmard jl, juliard jm, benamer h, himbert d, aubry p for the perm study group. efficacy of streptokinase, but not tissue-type plasminogen activator, in achieving 90-minutes patency after thrombolysis for acute myocardial infarction decreases with time to treatment. j am coll cardiol 1998; 31: 776-9.

[318] . Bosset a .j, vogt p., eeckhout e ., vanmelle g ., monnier p ., scaller m.d, stauffer j.c, kappenberger l, goy jj. long-term prognosis of patients having received thrombolysis in the acute stage of myocardial infarction. ann .cardiol , angéiol , 1997 , 46 (5-6) , 303 -310.

[319] . Thebault b, lefevre t, loiret j, bellorini m, guillard n, funck f. the management of myocardial infarction in patients aged 75 and over compared with that in patients aged 50 and over.

patients under 75. ann. cardiol.angéiol 1997; 46 (10): 635-641

[320] -Carney rj, murphy ga, brandt jr et a for the raami study investigators.randomised angiographic trial of recombinant tissue-type plasminogen activator (alteplase) in myocardial infarction. j am coll cardiol 1992; 20: 17-23.

[321] . Epidemiology of avoidable delay in the care of patients with acute myocardial infarction in italy. a gissi-generated study. gissi--avoidable delay study group. arch intern med. 24 jul 1995;155(14):1481-8.

[322] . Balbaa a, elguindy a, pericak d, natarajan mk, schwalm j. be

fore the door: comparing factors affecting symptom onset to first medical contact for stemi patients between a high and low-middle income country. int j cardiol heart vasc. 8 march 2022;39:100978.

[323] . Culic v. acute risk factors for myocardial infarction. int j cardiol. 25 apr 2007;117(2):260-9.

[324] . Pitts sr, niska rw, xu j, burt cw. national hospital ambulatory medical care survey: 2006 emergency department summary. natl health stat report. 6 august 2008;(7):1-38.

[325] . Goodacre s, cross e, arnold j, angelini k, capewell s, nicholl j. the health care burden of acute chest pain. heart. feb 2005;91(2):229-30.

[326] . Conti a, paladini b, toccafondi s, magazzini s, olivotto i, galassi f, et al. effectiveness of a multidisciplinary chest pain unit for the assessment of coronary syndromes and risk stratification in the florence area. am heart j. oct 2002;144(4):630-5.

[327] . Jh p, tp a, r r, rh w, ja f, jr b, et al. missed diagnoses of acute cardiac ischemia in the emergency department. the new england journal of medicine [internet]. 20 Apr 2000; 342 (16). available at: https://pubmed.ncbi.nlm.nih.gov/10770981/

[328] . Mccarthy bd, beshansky jr, d'agostino rb, selker hp. missed diagnoses of acute myocardial infarction in the emergency department: results from a multicenter study. ann emerg med. march 1993;22(3):579-82.

[329] . Silverman me. william heberden and some account of a disorder of the breast. clin cardiol. march 1987; 10(3):211-3.

[330] . Emergency department: rapid identification and treatment of patients with acute myocardial infarction. national heart attack alert program coordinating committee, 60 minutes to treatment working group. ann emerg med. feb 1994;23(2):311-29.

[331] . Gokhroo rk, ranwa bl, kishor k, priti k, ananthraj a, gupta s, et al. sweating: a specific predictor of st-segment elevation myocardial infarction among the symptoms of acute coronary syndrome: sweating in myocardial infarction (swimi) study group. clin cardiol. feb 2016;39(2):90-5.

[332] . Writing group members, mozaffarian d, benjamin ej, go as, arnett dk, blaha mj, et al. heart disease and stroke statistics-2016 update: a report from the american heart association. circulation. 26 jan 2016;133(4):e38-360.n

[333] . Canto jg, fincher c, kiefe ci, allison jj, li q, funkhouser e, et al. atypical presentations among medicare beneficiaries with unstable angina pectoris. am j cardiol. 1 august 2002;90(3):248-53.

[334] . Alexander kp, roe mt, chen ay, lytle bl, pollack cv, foody jm, et al. evolution in cardiovascular care for elderly patients with non-st-segment elevation acute coronary syndromes: results from the crusade national quality improvement initiative. j am coll cardiol. 18 oct 2005;46(8):1479-87.

[335] . Rosengren a, wallentin l, simoons m, gitt ak, behar s, battler a, et al. age, clinical presentation, and outcome of acute coronary syndromes in the euroheart acute coronary syndrome survey. eur heart j. apr 2006;27(7):789-95.

[336] . Donataccio mp, piadonataccio m, puymirat e, vassanelli c, blanchard d, le breton h, et al. presentation and revascularization patterns of patients admitted for acute coronary syndromes in france between 2004 and 2008 (from the national observational study of diagnostic and interventional cardiac catheterization [onaci]). am j cardiol. 15 jan 2014;113(2):243-8.

[337] . Kannel wb, abbott rd. incidence and prognosis of unrecognized myocardial infarction. an update on the framingham study. n engl j med. 1 nov 1984;311(18):1144-7.

[338] . Canto jg, shlipak mg, rogers wj, malmgren ja, frederick pd, lambrew ct, et al. prevalence, clinical characteristics, and mortality among patients with myocardial

infarction presenting without chest pain. jama. 28 june 2000;283(24):3223-9.
[339] . Brieger d, eagle ka, goodman sg, steg pg, budaj a, white k, et al. acute coronary syndromes without chest pain, an underdiagnosed and undertreated high-risk group: insights from the global registry of acute coronary events. chest. aoüt2004;126(2):461-9.
[340] . Braunwald e. the open-artery theory is alive and well--again. n engl j med. 25 nov 1993;329(22):1650-2.
[341] . Reimer ka, lowe je, rasmussen mm, jennings rb. the wavefront phenomenon of ischemic cell death. 1. myocardial infarct size vs duration of coronary occlusion in dogs. circulation. nov 1977;56(5):786-94.
[342] . Kloner ra, ellis sg, lange r, braunwald e. studies of experimental coronary artery reperfusion. effects on infarct size, myocardial function, biochemistry, ultrastructure and microvascular damage. circulation. aoüt 1983;68(2 pt 2):i8-15.
[343] . Gersh bj, stone gw, white hd, holmes dr. pharmacological facilitation of primary percutaneous coronary intervention for acute myocardial infarction: is the slope of the curve the shape of the future? jama. 23 feb 2005;293(8):979-86.
[344] . Gusto angiographic investigators. the effects of tissue plasminogen activator, streptokinase, or both on coronary-artery patency, ventricular function, and survival after acute myocardial infarction. n engl j med. 25 nov 1993;329(22):1615-22.
[345] . Acute coronary syndromes: an evidence-based review and... [internet]. relias media | online continuing medical education | relias media - continuing medical education publishing. available at: https://www.reliasmedia.com/articles/51388- acute- coronary-syndromes-an-evidence-based-review-and-outcome-optimizing- guidelines-for- patients-with-and-without-procedural-coronary-intervention-part-iii
[346] . Terkelsen cj, s0rensen jt, maeng m, jensen lo, tilsted hh, trautner s, et al. system delay and mortality among patients with stemi treated with primary percutaneous coronary intervention. jama. 18 august 2010;304(7):763-71.
[347] . Terkelsen cj, jensen lo, tilsted hh, trautner s, johnsen sp, vach w, et al. health care system delay and heart failure in patients with st-segment elevation myocardial infarction treated with primary percutaneous coronary intervention: follow-up of population-based medical registry data. ann intern med. 20 sept 2011;155(6):361-7.
[348] . Omer ma, tyler jm, henry td, garberich r, sharkey sw, schmidt cw, et al. clinical characteristics and outcomes of stemi patients with cardiogenic shock and cardiac arrest. jacc cardiovasc interv. May 25, 2020;13(10):1211-9.
[349] . Harjola vp, lassus j, sionis a, kober l, tarvasmäki t, spinar j, et al. clinical picture and risk prediction of short-term mortality in cardiogenic shock. eur j heart fail. may 2015;17(5):501-9.
[350] . Auffret v, cottin y, leurent g, gilard m, beer jc, zabalawi a, et al. predicting the development of in-hospital cardiogenic shock in patients with st-segment elevation myocardial infarction treated by primary percutaneous coronary intervention: the orbi risk score. eur heart j. 7 June 2018;39(22):2090-102.
[351] . Baran da, grines cl, bailey s, burkhoff d, hall sa, henry td, et al. scai clinical expert consensus statement on the classification of cardiogenic shock: this document was endorsed by the american college of cardiology (acc), the american heart association (aha), the society of critical care medicine (sccm), and the society of thoracic surgeons (sts) in april 2019. catheter cardiovasc interv. 1 Jul 2019;94(1):29-37.
[352] . Jentzer jc, van diepen s, barsness gw, henry td, menon v, rihal cs, et

al.Cardiogenic shock classification to predict mortality in the cardiac intensive care unit. j am coll cardiol. 29 oct 2019;74(17):2117-28.

[353] . Van diepen s, katz jn, albert nm, henry td, jacobs ak, kapur nk, et al. contemporary management of cardiogenic shock: a scientific statement from the american heart association. circulation. 17 oct 2017;136(16):e232-68.

[354] . Pöss j, köster j, fuernau g, eitel i, de waha s, ouarrak t, et al. risk stratification for patients in cardiogenic shock after acute myocardial infarction. j am coll cardiol. 18 Apr 2017;69(15):1913-20.

[355] . Farshid a, allada c, chandrasekhar j, marley p, mcgill d, o'connor s, et al. shorter ischaemic time and improved survival with pre-hospital stemi diagnosis and direct transfer for primary pci. heart lung circ. march 2015;24(3):234-40.

[356] . Otten am, maas ahem, ottervanger jp, kloosterman a, van 't hof awj, dambrink jhe, et al. is the difference in outcome between men and women treated by primary percutaneous coronary intervention age dependent? gender difference in stemi stratified on age. eur heart j acute cardiovasc care. dec 2013;2(4):334-41.

[357] . Van der meer mg, nathoe hm, van der graaf y, doevendans pa, appelman y. worse outcome in women with stemi: a systematic review of prognostic studies. eurj clin invest. feb 2015;45(2):226-35.

[358] . Kaul p, armstrong pw, sookram s, leung bk, brass n, welsh rc. temporal trends in patient and treatment delay among men and women presenting with st-elevation myocardial infarction. am heart j. jan 2011;161(1):91-7.

[359] . Ladwig kh, meisinger c, hymer h, wolf k, heier m, von scheidt w, et al. sex and age specific time patterns and long term time trends of pre-hospital delay of patients presenting with acute st-segment elevation myocardial infarction. int j cardiol. 3 Nov 2011;152(3):350-5.

[360] . Ladwig kh, meisinger c, hymer h, wolf k, heier m, von scheidt w, et al. sex and age specific time patterns and long term time trends of pre-hospital delay of patients presenting with acute st-segment elevation myocardial infarction. int j cardiol. 3 Nov 2011;152(3):350-5.

[361] . Borowicz a, nadolny k, bujak k, ciesla d, gasior m, hudzik b. paramedic versus physician-staffed ambulances and prehospital delays in the management of patients with st-segment elevation myocardial infarction. cardiol j. 25 feb 2021;28(1):110-7.

[362] . Kubica j, adamski p, ladny jr, kazmierczak j, fabiszak t, filipiak kj, et al. prehospital treatment of patients with acute coronary syndrome: recommendations for medical emergency teams. expert position update 2022. cardiol j. 2022;29(4):540-52.

[363] . Kubica j, adamski p, paciorek p, ladny jr, kalarus z, banasiak w, et al. antiaggregation therapy in patients with acute coronary syndrome - recommendations for medical emergency teams. experts' standpoint. kardiol pol. 2017;75(4):399-408.

[364] . Kubica j, adamski p, paciorek p, ladny jr, kalarus z, banasiak w, et al. treatment of patients with acute coronary syndrome: recommendations for medical emergency teams: focus on antiplatelet therapies. updated experts' standpoint. cardiol j. 2018;25(3):291-300.

APPENDIX I: Study clinical form

ANNEX II :

Calculation of TIMI risk for ST+ ACS from Morrow DA et al [113].

ANNEX III :

Treatment algorithm for tachycardia in patients with ACS

ANNEX IV :

Appendix IV: Patient presentation methods, components of ischaemic duration and reperfusion strategy algorithm. ESC 201

Annexel CHU Mostaganem Service des Urgences le : |_|_|/|_|_|/|_|_|

ACS with ST elevation

ADMINISTRATIVE DATA :

□ Fiche N*: Sexe : F □M □

□ Patient surname : Patient's first name :

□ Age: Weight: Height :

□ Address:

□ Telephone (patient): Informed consent **Socio-economic level: good □medium □ poor □**

MEDICAL DATA AND RISK FACTORS :

□ Familial inheritance of cardiovascular disease (<65 years): ouianon □

□ Active smoking yes□ No

□ Diabetes yes □ no □

□

□ Status: alive □ deceased □lost to sight : □

□ HTA yes□ no □

□ Renal insufficiency yes□ no □

□ Obliterative arteritis of the lower limbs yes□ no □

□ Previous ischemic heart disease yes□ no □

□ Ischaemic stroke yes□ no □

□ Haemorrhagic stroke yes□ no □

□ Heart failure yes□ no □

Π Dyslipidaemia yes□ no □

□ Sedentary lifestyle yes □ no □

□ Allergies yes □ no □

□ Commentary:

□ Other medical history :

□ Surgical history :

□ Treatment in progress :

POSITIVE DIAGNOSIS :

Start of symptomatology: |_|_|/|_|_|/|_|_|; **Time**: Hmin

□ Initial symptom: chest pain □ lipothymia □ epigastralgia □ dyspnoea □ palpitation □ syncope □ RTA □ Other □

□ Consultation time : ; Consultation pain time : H Min

Initial clinical examination:

□ **CGS=** || |/15, PAS/PAD= |_|_|_|mm Hg, **HR=** bpm ; SpO2= %.

□ **Finger blood glucose**: ||| ,||| g/l

□ **Initial KILLIP stage**: 1 □2 □3 □4 □

□ **Signs of acute circulatory failure:** yes□ noa

□ **DCI sign:** .... yes□ noa

□ **ACR**: ouia nona

18-lead ECG :

□ **Repolarization disorder: ST+ supershift** : Anteroseptal aAnteroseptо-apical α
Extended anterior a Deep Inferior a Lateral Top □ Basal □
Latéro Basal a Septal a Sub-basal Infero Lateral
Infero-latero-basal a Circumferential a Low lateral a

□ **Conduction disorder**: yes a no a

BBG : yes no a

□ **Rhythm disorder**: yes no a

If so, which one?

Final diagnosis :

ACS ST+ a ACS non-ST+ a Other diagnosis :

TREATMENT: 1. Primary angioplasty,2. Thrombolysis

I I

Emergency thrombolysis: 1. pre-hospital by EMS, 2. hospital in UMC
I I

- Start date : | ||/| ||/|| |; Time|- ||- ||- |Min
- Molecule: **Metalysis** □ **Actylise** □
- If no, **Reason**: Spontaneous reperfusion □ Delay considered too late □Patient at risk

□
Thrombolysis contraindication □______Diagnostic doubt □ Other □ :

Other emergency adjuvant treatments:

Aspirin (□	Clopidogrel ( ) °	HNF ( ) □
LMWH (□	Morphine ( ) °	Nitro derivatives (...)o
Beta blocker (..) °	Cordarone ( ) °	Lasilix() □
Atropine ( ) °	Xylocaine ( ) °	EEC ( ) □
RCP ( ) °	Cathecolamines ;	Dose

Other □ :

Progression during hospitalisation in the UMC :

Simple □ Ischemic recurrence □
Rhythmic complication □ (excluding RIVA); Conductive disorder □
EDC □ Heart failure □
Haemorrhage: Major □ Minor □ (Seat)
STROKE : Ischemic □ Hemorrhagic □
ACR □ Other □ :

□ **Death during hospitalisation:**yes □ no □
If yes, Date ||_ |/|| ИU- | ; Time || |H| || Min
□ **Emergency discharge :** **Date** | ||/|||/| ||| ; Time||H| ||Min
□ **Length of stay in emergency ICU :** **d h**
Destination: return home □ interventional cardiology □ yes □ no □

Developments during hospitalisation:

Cardiac echo: LVEF :
□ **Complications:** Shock □ Acute heart failure □
□ Rhythm disorders (excluding RIVA) □ : Disorder conductive □ :
□ Hemorrhagic: major □ Minor (Seat)
□ Stroke □: Ischaemic □ Haemorrhagic □
ACR □ Other □:
□ Death during hospitalisation: yes □ no □
If yes, date and time: || |/|||/|||| "Kernel "¡f ήH| || |Min

Definitive diagnosis :

ACS ST+ □ ACS no ST+ □ Other diagnosis □:

Leaving the hospital:

Date and time: | ||/| ||/| ||| ; Time |||H|| |Min Destination: back
Home □ Return home □ Intensive care □ Surgery □

BECOME 30 DAYS

□ Therapeutic compliance: Good □Poor □
П Alive without complications: yes □ no □
If complications:
Angina □Dyspnoea □Chronic heart failure (LVEF=) □
SCA St +□ ACS no ST++ □ sent thrombosis □
□ Haemorrhagic syndrome yes □ no □
If yes, specify location :
□ Chronic heart failure: yes □ no □

□ Death: yes □ no □

If yes, cause of death :

Appendix II: Calculation of TIMI risk for ST+ ACS based on Morrow DA et al [113].

Risk factors	**Number of points**
Age 65-74 / > 75	**2 / 3**
Systolic blood pressure < 100	**3**
Heart rate > 100	**2**
Kilip II-IV	**2**
Anterior Sus-ST or LBB	**1**
Diabetes, hypertension or angina	**1**
Weight < 67 kgs	**1**
Time between pain and treatment > 4 hours	**1**
Calculation of the TIMI ST+ risk score (cumulative points)	**Mortality at 30 days (% of total)**
0	**0.8**
1	**1.6**
2	**2.2**
3	**4.4**
4	**7.3**
5	**12.4**
6	**16.1**
7	**23.4**
8	**26.8**
> 8	**35.9**

Appendix III: Treatment algorithm for tachycardia in patients with ACS

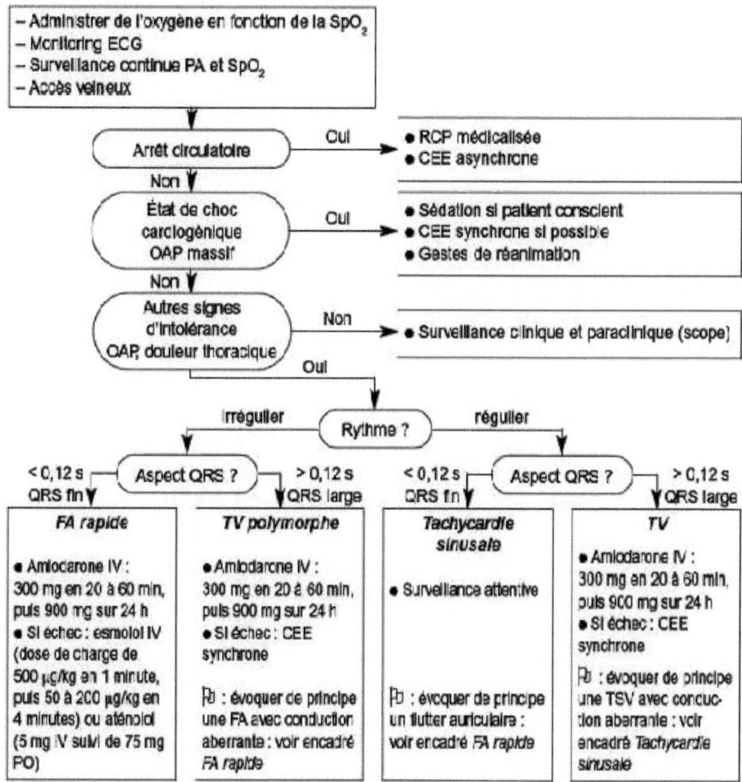

Algorithm 4. Treatment algorithm for tachycardia (heart rate > 100 bpm) in a patient with ACS, excluding vital failure and well-tolerated tachycardia (after J.E. de la Coussaye et al}.

Appendix IV: Methods of patient presentation, components of ischaemic duration and reperfusion strategy algorithm. **ESC 201**

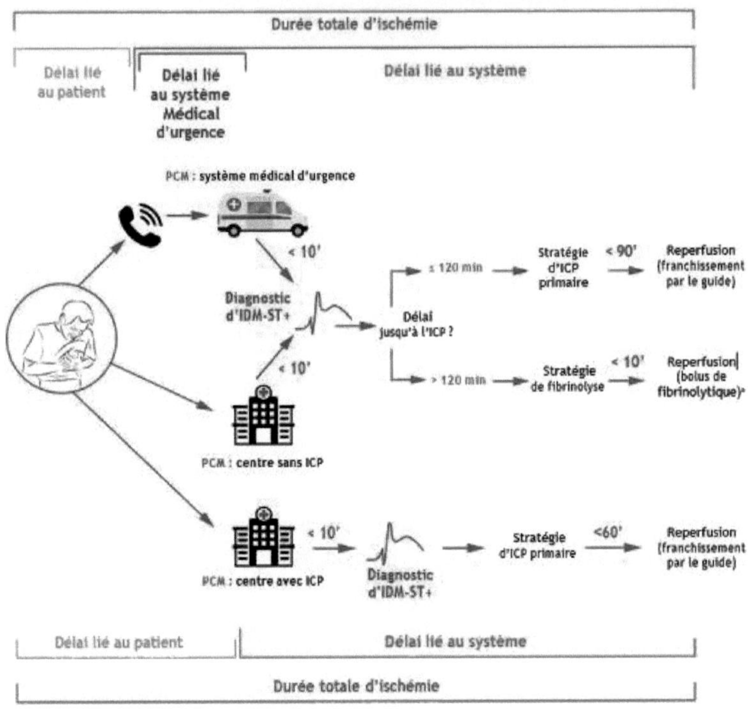

Summary

Introduction :

The management of ST-segment elevation acute coronary syndrome (STEMI) is still a problem in regions outside the country's major metropolitan areas.

The aim of our study is to demonstrate the value of setting up a thrombolysis network for STEMI patients in the Mostaganem region.

Methods :

Single-centre prospective observational study, carried out from 1 December 2018 to 30 November 2022, including STEMI patients undergoing thrombolysis reperfusion managed as part of a new thrombolysis network in Mostaganem. A descriptive analysis was performed on the course, characteristics, management and intra-hospital prognosis of these patients.

Results :

Of the 10 STEMI patients who underwent thrombolysis in 2018 before the network was set up, an average of 74 patients underwent thrombolysis each year as part of the Mostaganem thrombolysis network.

We included 295 patients, with an average age of 58 years and a predominance of men (83.7%); the average pain-hospitalisation time was 3 hours 42 minutes (2 hours 48 minutes for the city of Mostaganem vs. 4 hours 21 minutes for the surrounding area; $p<10^{-3}$); we recorded 74.6% cases of successful reperfusion, with 19.7% complications and 6.7% mortality (6.4% cardiovascular and 0.3% due to serious haemorrhage).

Multivariate analysis enabled us to identify independent predictive factors for the occurrence of early complications: age > 65 years (ORa 2.6 [1.0-6.5]; p= 0.043), origin (ORa 4.1 [1.6-10.7];p= 0.004), heart rate on admission >100/min (ORa 8.6 [3.2-22.9] ;p< 10-3) and thrombolysis failure (ORa 10.0 [4.025.0] ;$p< 10^{-3}$) and mortality: cardiogenic shock (ORa 299.2 [35.2-2542.4]; $p < 10^{-3}$) and thrombolysis delay > 6H (ORa 6.2 [1.1-35.6]; p=0.042).

Conclusion:

The installation of a thrombolysis network in Mostaganem has made it possible to significantly improve the medical management of STEMI by reducing delays and thus offering the opportunity of medicated reperfusion to a larger proportion of patients who will be able to treat STEMI effectively.

Key words: ST+ acute coronary syndrome, network, thrombolysis, delays, Mostaganem.

Printed by Books on Demand GmbH, Norderstedt / Germany